P9-AFZ-425

# HUMAN HEALTH AND THE ENVIRONMENT
## A Turn-of-the-Century Perspective

R

# HUMAN HEALTH AND THE ENVIRONMENT
## A Turn-of-the-Century Perspective

by


**Donald Vesley**
*Division of Environmental and Occupational Health*
*School of Public Health*
*University of Minnesota*

**KLUWER ACADEMIC PUBLISHERS**
**Boston / Dordrecht / London**

**uth America:**

Assinippi Park
Norwell, Massachusetts 02061 USA
Telephone (781) 871-6600
Fax (781) 871-6528
E-Mail <kluwer@wkap.com>

**Distributors for all other countries:**
Kluwer Academic Publishers Group
Distribution Centre
Post Office Box 322
3300 AH Dordrecht, THE NETHERLANDS
Telephone 31 78 6392 392
Fax 31 78 6546 474
E-Mail <services@wkap.nl>

Electronic Services <http://www.wkap.nl>

---

**Library of Congress Cataloging-in-Publication Data**

Vesley, Donald.
Human Health and the environment : a turn-of-the-century perspective / by Donald Vesley.
p. cm.
Includes bibliographical references and index.
ISBN 0-7923-8616-7 (alk. paper)
1. Environmental health. I. Title.
RA565.V47 1999
615.9'02--dc21 99-37324
CIP

---



*Printed on acid-free paper.*

***Original cover art provided by Catherine Vesley.***

Printed in the United States of America

## CONTENTS

**Human Health and the Environment: A Turn-of-the-Century Perspective**

# PREFACE

The twentieth century has seen a remarkable evolution of environmental health and environmental protection concerns and concepts in the United States. As a teacher of Environmental Health since the late 1950s, I have witnessed the many twists and turns that have marked the latter half of the century, and have had to seek continuously to explain these phenomena to my students in some rational manner. We have witnessed the following and more: great progress in controlling acute infectious diseases through successes in drinking water treatment and food processing; the emergence of greater concern with trace chemicals in air and water and their role in chronic disease causation; conflicting attitudes toward miraculous chemicals such as DDT (which promised relief from arthropod-borne scourges, then came to be seen as another chemical threat to our children and our environment); then the reemerging concern with infectious diseases precipitated by blood-borne pathogens such as HIV. All this occurred against a backdrop of scientific uncertainty and amid failures of risk assessment and risk communication, together with press sensationalism—from "mad cow disease" to "flesh eating" streptococci. No wonder the public is confused.

Now, as we enter a new century, I would like to summarize this important and all-encompassing field for a wide audience. Although the book is written from the perspective of our technologically advanced nation, I will address throughout the very different environment, and different human health problems arising from the environment, in third world and emerging societies. The incredible advances in transportation and rapidly integrating and entwining world economies have turned previously isolated disease outbreaks into potential pandemics that require increasing vigilance and surveillance to prevent widespread catastrophes. The emerging and reemerging infections that have been so widely publicized in recent years may or may not be overblown. However, there is no question that the safety and sanitary quality of American life cannot be taken for granted. We cannot afford to relax our efforts or compromise the public health infrastructure built on so many hard-won advances during the twentieth century. By no means do I expect to lay out an unequivocal solution to the many dilemmas and paradoxes which characterize environmental health at the end of the twentieth century. I attempt only to summarize what I see as the issues and relationships among innumerable contrasting viewpoints. I challenge the reader to consider the issues and draw his or her own conclusions, striking a balance between science, politics, and societal values that may lead to the optimal "net societal benefit"—which we all seek but which remains frustratingly elusive.

## ACKNOWLEDGMENTS

Numerous individuals have influenced my thinking about environmental health issues during the many years that this book was incubating as I taught a survey course to both undergraduate and graduate students at the University of Minnesota. The following have been particularly influential, but I also acknowledge the many others who are too numerous to mention but who do not go unappreciated.

Specific acknowledgments are offered to Professor Ted Olsen, whose course in Sanitary Biology sparked my enthusiasm for the field; to my initial advisor, the late Professor Herbert Bosch, and my subsequent mentors, the late professors Richard Bond and George Michaelsen, who gave me the opportunity to teach and do research in Environmental Health; to Professor Velvl Greene, who served as mentor and colleague in early research and teaching efforts and has provided career-long friendship and support even when we disagreed on the issues; to Mary Halbert, whose dedication to excellence in the laboratory and to teaching students how to "do it right" kept me on my toes at all times; to Melissa Nellis, former student and current colleague, whose loyalty and support in teaching and research efforts has kept me going through good times and bad; to all of the dedicated teaching assistants whose help has been unwavering over the years, but particularly to my recent "T.A.s", Nicole Vars McCullough, whose enthusiasm for the teaching process convinced me to modernize my methods, and Teresa Kurtz, who has stuck with me for the past two years, renewed my faith in the younger generation, and inspired my optimism for the future.

I extend my appreciation to my talented daughter-in-law Kristine and to my son Mark for their copy editing and indexing skills and to Karen Brademeyer for her word processing assistance. I thank my wife, Catherine, for her love and patience throughout the project. Finally, I dedicate the book to the hundreds of students who have endured patiently through the course over the years. I hope they have learned and remembered at least something from it.

# Section I

# INTRODUCTION

# Chapter 1

# HUMAN HEALTH AND THE ENVIRONMENT AS THE TWENTIETH CENTURY ENDS

> With acid rain, ozone depletion, chemical pollution, radioactivity, the razing of tropical forests and a dozen other assaults on the environment, we are pushing and pulling our little world in poorly understood directions.
>
> — Carl Sagan, *Billions and Billions*

The relationship between human health and the physical environment is both obvious and obscure. It is obvious in the case of acute microbial agents or chemical toxins when symptoms occur soon after ingestion, inhalation, or absorption. It is obscure in the case of long-term effects of trace contaminants that produce no acute symptoms but are suspected of involvement in chronic disease by virtue of extrapolation from *in vitro* or *in vivo* laboratory studies or epidemiological investigations. It should not surprise us that scientists are not always in agreement as to the magnitude or even the nature of environmental risk (indeed, we cannot even agree definitively on whether the twentieth century ends on December 31, 1999 or on December 31, 2000). Nor should we be surprised that environmental standards and regulations are often deemed too strict by industry and conservatives while at the same time too lax by environmentalists and liberals.

Clearly, in the industrialized world, there has been progress in making the population healthier through environmental controls since the start of the twentieth century. The great successes of environmental health, such as filtration and disinfection of drinking water and pasteurization of dairy products, are unequivocal and have contributed to significant decreases in many enteric diseases. Unfortunately, most of the world's population still lives in developing nations where populations burgeon and resources available for infrastructure are inadequate to provide such basic protection. A major challenge for the twenty-first century will be to provide cleaner water, air, and food to more people. Many children still die from ingestion of contaminated drinking water or inhalation of particulates from indoor cooking fires. Clearly, significant improvement in environmental health can only occur in the context of improving economic development (which in turn creates additional pollution) and accompanying population control (which is fraught with social and religious obstacles). Thus, the challenge for environmental health professionals in the next century is daunting.

Meanwhile, the major environmental debates in the United States and other industrialized nations focus on issues of little current interest to the developing world. Reducing greenhouse gasses to prevent global warming and phasing out chlorinated fluorocarbons (CFCs) to preserve the ozone layer simply are not important in countries which cannot afford significant quantities of fossil fuels (although they continue to burn forests to clear land for agriculture) or refrigerants containing CFCs. Similarly, in third world countries, more judicious use of pesticides and greater reliance on integrated pest management are less important than the need to grow more food and to prevent vector-borne disease (that transmitted via arthropods or other animals). The political difficulty in obtaining global cooperation for environmental protection is not surprising in light of such economic inequity.

How do we even begin to resolve the discrepancies and disagreements between conservatives and liberals in the industrialized world and between the haves and have-nots in the world community? The ideal solution would be a magic formula that would calculate something called *net societal benefit*. If such a formula existed, we could calculate the optimal approach to standards and regulations that would do the most good (health and prosperity) for the greatest number of people. Unfortunately, such a formula does not exist. Thus, we are left with the ongoing tug of war between competing interests and the inevitable compromises, which may or may not end up being in anyone's best interest. Ideally, increasing scientific knowledge, sharpened tools of risk assessment, and improved skills in risk communication will serve to narrow the gulf of conflicting perceptions and bring us closer to that ideal of net societal benefit.

While human ingenuity and civilized dialogue are indeed the potential positives in ultimate resolution of environmental issues, other human factors continue to foster ongoing conflict in many parts of the world and promote the very real threat of terroristic tactics (including environmental sabotage and nuclear, chemical, and biological attacks). The other great force to be reckoned with is nature. The continuing emergence of exotic infectious diseases serves to remind us of the substantial evolutionary head start that microbes had over humans. Rapid evolution of microbes is fueled by their rapid generation time (minutes instead of years under ideal conditions). Examples include the influenza virus adapting itself to direct transfer from poultry to humans, and the prions of bovine spongiform encephalopathy (BSE) in the form of new variant Creutzfeldt-Jakob disease (nv CJD) being apparently transmissible to humans through ingestion of beef. These examples serve to remind us of our continued susceptibility to the foibles of nature. In addition, human folly can serve to give these natural events added impetus. Feeding neurological tissue to cattle was instrumental in spreading BSE, and continued infringement of human populations on jungles and rain forests has exposed humans to agents that may then cross species barriers—the emergence of AIDS in the 1980s was explained as arising from exposure to chimpanzee blood during butchering of the animal for food.

In the ensuing chapters I will attempt to make some sense of the array of environmental health issues facing the United States and the world in the coming century (and the new millennium). I will define the environment broadly and attempt to identify the importance of environmental factors in the scope of human health effects. I will discuss sources of hazardous physical, chemical, and biological agents, pathways to the human host, and approaches to environmental intervention. All of these concepts are necessary if the general public is ever to understand and prioritize the issues. Most of us are informed of threats to our health through sensationalized reporting in the press and on television. Effective risk communication has been a major failure of public health professionals. Thus, public fear of toxins, carcinogens, pathogens, and anything vaguely radioactive without the perspective of dose or exposure potential remains to be assuaged.

As environmental health professionals, we cannot ignore the public perceptions regardless of scientific merit. The "outrage factor" is very real and refers to the public concern about environmental forces over which they have no control. Thus, opposition to municipal incineration of solid waste, reluctance to accept irradiated food, and even prevention of fluoridation of drinking water have stymied advances in public health. Meanwhile, voluntary behaviors, such as alcohol and tobacco use and failure to wear seat belts, with much greater potential for health destruction, continue to flourish. This raises the very important issue of responsibility, again fraught with political overtones: we must come to grips with the balance between individual and government responsibility (with some measure of corporate responsibility thrown in). Nowhere has this balance been more publicized than in the ongoing saga of tobacco products and human health. The issue remains an enigma to the public health profession in that a product responsible for more than 400,000 deaths annually in the United States continues to be marketed legally. In contrast, other products are banned because of a suspected link to cancer in the magnitude as small as one excess case per 1,000,000 population. While the tobacco companies plead that this is a matter of personal choice (read: individual responsibility), the government struggles to find legislative approaches just to keep this deadly product out of the hands of children too young to make such a choice. The influence of political campaign contributions from the tobacco companies highlights the need for reform of the political system to reduce the impact of such contributions and to assure that the health of our children (and their future health as adults) becomes a more important national priority.

## MAJOR ENVIRONMENTAL HEALTH ISSUES AT THE TURN OF THE CENTURY

What are the major challenges facing the environmental health profession at the turn of the century? In order to answer such a question we must resolve other philosophical questions concerning our definitions of *human health* and *optimum health* as well as resolving our uncertainties about the relationship between health and environment. Unfortunately, it is easy to get bogged down in semantics and scientific limitations, which could paralyze any initiative to address important issues. Thus, recognizing the lack of consensus, we must make some assumptions as a basis for prioritization.

First, we must realize that humans are part of the natural world, subject to the whims of nature, which we simply cannot always control. Cycles of warming and cooling, flood and draught, devastating earthquakes and storms, feast and famine have occurred throughout human history and are only marginally affected by human activity. Second, we must recognize that the problems of industrialized nations do not coincide with the problems of developing nations and that political resolution of those differences must be a part of any global strategy for environmental control. If we assume that quality of life is a worthwhile goal, then population control must be an integral part of future progress. The depletion of natural resources and pollution associated with burning of fossil fuels are serious enough when the majority of humans contribute little to this depletion. The consequences would be unsustainable if development in emerging societies progressed along the same path. The conflict is clear in recent attempts at imposing global limits on greenhouse gasses, as third world nations view such limits as unfair impediments to development.

Following are a few of the major issues that I see dominating the environmental health profession as we enter the twenty-first century (not necessarily in order of importance):

## Nuclear Holocaust and Other Forms of Terrorism

The capacity of humans for self-destruction clearly poses an enormous threat to civilization as we know it. Throughout the cold war years we lived under the threat of nuclear war and the recognition that a "nuclear winter" could potentially make the planet uninhabitable for centuries to come. The mutual deterrence philosophy of the nuclear powers (and perhaps a little old-fashioned good luck) protected us from that calamity for the duration of the cold war.

The new threat comes from nuclear proliferation (as marked by the Indian and Pakistani tests in 1998) and the potential for nuclear capacity to be acquired by rogue nations or even independent terrorist groups. While the current threat may no longer have the global implications of the cold war days, it remains a real concern with little hope for completely stuffing the genie back into the bottle. In addition to nuclear terrorism, chemical or biological terrorism will remain as a major concern into the twenty-first century. The release of chemical or biological agents into the environment of densely populated buildings, airports, or subway systems is a virtually untested possibility. The sarin attack in 1995 in the Tokyo subway serves as a warning that such events can happen with serious health consequences locally, if not the more global horrors of nuclear holocaust.

## Global Warming

Few environmental issues are receiving as much attention as the potential for human activity to alter global climate, unleashing enormous ecological and health effects. There is far from a consensus as to the imminence or seriousness of the warming trend. Scenarios for dire consequences abound. The worst-case scenarios include melting of the polar ice caps with flooding of coastal cities, and devastating declines in crop production resulting in mass starvation. Expanded ranges of disease vectors and extinction of species with unknown effects on the food chain can also be postulated. Many scientists believe that we cannot wait for more definitive evidence that human industrialization and the concomitant release of greenhouse gasses from burning fossil fuels is fueling the increase in $CO_2$ most responsible for the warming trend. Others believe that such effects are balanced by increased growth of green plants, which utilize the $CO_2$ to produce more oxygen. This issue is severely clouded by the economic implications for industrialized nations and the desire of developing nations to industrialize. The middle course of seeking alternative energy sources and imposing some restrictions on emissions will probably prevail at least until more definitive evidence is at hand that the climate is indeed warming and that we are not in fact headed toward another ice age.

## Depletion of the Ozone Layer

Of all the major environmental controversies, the depletion of the ozone layer, which is needed to filter out ultraviolet rays from the sun, seems to be nearest to a scientific consensus. Although the magnitude and permanence of ozone layer depletion may not be clear, both the existence of ozone holes and the chemical reaction of chlorofluorocarbons (CFCs), which contributes to ozone depletion, have been established. Human health effects in the form of skin cancer and cataracts caused by ultraviolet rays are also well documented. There are hopeful signs that manufacture and use of the refrigerants containing chlorofluorocarbons have been greatly reduced and that substitute refrigerants are now available. It remains to be seen how effective these measures will prove to be, but at least for this problem the prospects appear to be hopeful.

## Chemicals and Cancer

There are few issues that engender more public concern than the potential for involuntary exposure to chemicals that cause cancer. As discussed earlier in this chapter, our successes in environmental health in the twentieth century have come largely in acute infectious disease control through such advances as chlorination of drinking water and pasteurization of dairy products. The association of those vehicles (food and water) with acute infectious disease was definitive, and the effectiveness of the control measures was clearly demonstrated.

We have no such clarity in our attempts to define the role of trace chemicals over a long period of time in the causation of human cancer. Thus, the process of defining risk and setting standards is more convoluted and uncertain. The example of chlorine in drinking water illustrates the difficulty of this task as well as the complex interplay of problems and solutions. The advantages of chlorination in preventing enteric disease are well accepted, but this success is now taken for granted, and public concern has shifted toward the potential role of chlorine in cancer causation. Chlorine reacts with organic matter to form trihalomethanes (typically chloroform). Chloroform, in turn, has been shown to be a weak carcinogen in some rodent species. The EPA has thus set an upper limit on trihalomethanes in drinking water (50 ppb), prompting changes in chlorination practice to comply with the limit. This, in turn, raises concerns that effectiveness in preventing enteric disease could be compromised and becomes a classic example of "net societal benefit," with the number of excess cancer cases resulting from chlorination being the great unknown in the equation. The complex interaction of trace chemicals with the ecology of the environment is highlighted by the current worldwide dilemma of decreasing frog populations and increasing reports of frog deformities. Trace pesticides are suspected of being a causative factor, but proof is elusive, and possible implications for human health (particularly related to cancer) remain uncertain.

## Emerging Infectious Diseases

One other environmental health issue that certainly will occupy increasing attention in the twenty-first century is that of emerging infectious diseases. While the attention of the industrialized world has turned from infectious to chronic diseases with advances in public health and medicine, more than 10 percent of the world's population is afflicted with infectious diseases such as malaria, schistosomiasis, and tuberculosis. Those diseases remain by far the most prominent debilitating conditions for hundreds of millions of the world population.

Meanwhile, AIDS has become the prime example of a new pandemic afflicting millions of people worldwide with a disease that takes years to run its course; 90 percent of those afflicted do not know that they have been infected and thus do not seek early treatment. Increasing antibiotic resistance, changing vector habitats, and human encroachment into rain forests and jungles all contribute to this emergence, raising major issues of resource allocation for research and prevention. The worldwide emergence of new and reemerging infectious disease also raises the issue of rapid spread from continent to continent with modern jet transportation allowing infected but prodromal individuals to travel anywhere in the world before they become symptomatic. Many of the important emerging infectious diseases and their global implications will be discussed in subsequent chapters.

## REFERENCES

The following textbooks and reference books are among the more recent summaries of environmental health issues and information. All of these have provided valuable insights for my synthesis of the subject.

### TEXTBOOKS

Blumenthal, D. S. and A. J. Ruttenber. 1995. *Introduction to Environmental Health* (2nd Ed.). Springer Publishing Co. New York.

Moeller, D. W. 1997. *Environmental Health* (Rev. Ed.). Harvard University Press. Cambridge, MA.

Morgan, M. T. 1997. *Environmental Health* (2nd Ed.). Brown and Benchmark Publishers. Dubuque, IA.

Nadakavukaren, A. 1995. *Our Global Environment: A Health Perspective* (4th Ed.). Waveland Press. Prospect Heights, IL.

ReVelle, P. and C. ReVelle. 1992. *The Global Environment.* Jones and Bartlett Publishers. Boston.

Wildavsky, A. 1995. *But Is It True? A Citizen's Guide to Environmental Health and Safety Issues.* Harvard University Press. Cambridge, MA.

### REFERENCE BOOKS

Salvato, J. A. 1992. *Environmental Engineering and Sanitation* (4th Ed.). John Wiley & Sons. New York.

Koren, H. and M. Bisesi. 1996. *Handbook of Environmental Health and Safety* (3rd Ed.). Vols. 1 and 2, CRC, Lewis Publishers. Boca Raton, FL.

# Chapter 2

# EPIDEMIOLOGY AND ENVIRONMENTAL HEALTH

In order to establish effective environmental health programs, locally, nationally, or globally, we must have some basis for resource allocation that will optimize those programs to the greatest health benefit for the most people. Unfortunately, such allocation faces seemingly insurmountable obstacles in the form of political reality and scientific limitations. Nonetheless, there must be links between epidemiological knowledge, surveillance to determine the magnitude of problems, and ultimate environmental (or public health) intervention strategies. The foundation of public health is the concept that it is preferable (in both the personal and the societal realms) to prevent illness rather than to depend on cures. Thus, Public Health has always encountered a visibility problem in that it is most successful when disease is not occurring and therefore not newsworthy for the mass media.

What we must do is learn as much as possible about each disease and maintain surveillance over incidence and prevalence trends so that preventive strategies can be instituted as quickly as possible. To do this we must depend on the practice of epidemiology. The epidemiology of a human disease, in the broadest sense, is the study of those factors which influence the occurrence of the disease in populations as opposed to individuals.

## EPIDEMIOLOGICAL TERMINOLOGY

### Endemic

The endemic level of a disease refers to the number of cases expected to occur in a given community in a given period (usually a year). This level may remain relatively constant over a long period of time as in the case of the common cold, or it may show a downward trend when public health measures (and/or medical advances) are successful as in the case of the declines in tuberculosis and polio in the U.S. in the second half of the twentieth century. It is also apparently possible to eradicate a disease completely, even on a worldwide basis—the eradication of smallpox was brought about by a vaccine that reduced the reservoir of infected humans to a level below which the disease could not sustain itself. It is expected that the twenty-first century will see more such successes with several diseases, including polio and measles, being likely candidates for eradication. Endemic trends can also be upward, signaling the need for additional preventive measures. Dengue fever in the Caribbean and South Asia in the 1990s is an example of such a rising trend.

**Epidemic**

An epidemic is the occurrence of disease in numbers significantly greater than expected. The number of cases can be very large, such as in influenza outbreaks, where the seasonal endemic level is usually high to begin with, or very small, such as for human rabies, where the endemic level in most countries is close to zero. Surveillance is critical for early detection and control of epidemics.

**Pandemic**

Pandemics are epidemics that spread to many different areas of the world. Historically, the most publicized of these were the plague pandemics of the Dark Ages, including the "Black Death" of the fourteenth century, which is said to have killed one-third of the population of Europe. The Black Death was one of the very few events in recorded human history that actually caused a downturn in the exponential increase in human population. The 1918 influenza pandemic is another example; this outbreak killed millions of people worldwide just a few months before abating and mysteriously fading away almost as quickly. The AIDS pandemic of the late twentieth century is the most recent example of a pandemic and a classic example of an emerging infectious disease. First detected and publicized in the United States in 1981 (although now known to be of African origin), it is spreading most dramatically in Africa and South Asia, with significant impact on resources for treatment and research, possibly at the expense of other human health problems. This situation is exacerbated by the lengthy course and expensive treatment regimens associated with the disease. The search for a vaccine continues but as of 1999 remains elusive.

**Zoonosis**

Zoonotic diseases are those transmitted to humans from other animal species. There are many examples of such diseases, including rabies, anthrax, and plague. They involve various routes of transmission and sometimes arthropod vectors as well, thus requiring many varied intervention strategies.

### Disease Incidence

*Incidence* is the number of new cases of a given disease in a given population during a given time period (usually one year). It is used primarily for acute illnesses.

### Disease Prevalence

*Prevalence* is the number of cases of a given disease at a particular point in time. It is used primarily for chronic illnesses such as multiple sclerosis or muscular dystrophy, where people may continue to live with the condition for long periods.

### Common Source Outbreaks

This refers to incidents where multiple individuals acquire an acute illness from a common single source or event. Prominent examples are food-borne or waterborne outbreaks, which are of particular importance in environmental health.

### Secondary Transmission

Secondary transmission refers to those diseases that are acquired from a primary source and can then be transmitted from an infected individual directly to another person. Many enteric diseases such as hepatitis A and typhoid fever are in this category. Other diseases, such as legionellosis and malaria, have not been shown to be transmitted secondarily.

### Types of Agents

The agents of human illness can be categorized as *infectious* (including bacteria, viruses, fungi, protozoa, and helminths), *carcinogenic* (cancer causing), *teratogenic* (causing harm to an unborn fetus), or *toxic*. The carcinogenic and teratogenic agents are primarily chemical, although physical agents such as ionizing and nonionizing radiation are also applicable. They have been particularly difficult to deal with because the effects are long-term and chronic rather than short-term and acute, a theme which will recur frequently in this book. In addition, there are the many agents of trauma (from guns to automobiles to unguarded machinery) which are under the province of environmental health and subject to environmental intervention as I will describe it in this book.

## THE EPIDEMIOLOGICAL CHAIN

Although applied primarily to infectious diseases, the concept of *epidemiological chain* can be adapted to other agents of human illness. Understanding the various links is imperative to seeking the weakest link to interrupt the chain and prevent illness.

### The Etiological Agent

The various types of biological, chemical, and physical agents have been described above. This link is not generally amenable to direct intervention except in the sense that we can avoid the use of certain natural or xenobiotic (man-made) chemicals by substituting innately less hazardous substances for particular uses. That concept I will refer to as *prevention*, a currently very popular theme in environmental health which will recur throughout this book. The concept of prevention is often contrasted with that of "end-of-pipe" solutions in which we intercept the agents at some point on the pathway to human exposure.

### The Reservoir

The natural habitat of an agent is referred to as its *reservoir*. This applies primarily to infectious agents, which must have a natural home base for self-perpetuation. Reservoirs can include humans and other animals (feral or domestic), soil (containing, for example, botulism or tetanus spores), or water (containing, for example, *Legionella* bacteria or some *Vibrio* species which inhabit ocean sediments). The reservoir is often not amenable to direct intervention except in the case of certain vermin such as domestic rats, which serve as reservoirs for plague and murine typhus.

### Escape or Release from the Reservoir

For human or other animal sources, escape from the reservoir can be through natural respiratory, enteric, or urinary excretions. For blood-borne pathogens, it can be through the bite of an arthropod vector. Soil organisms can contaminate crops and water organisms can be released through aerosolization (*Legionella*) or uptake in the food chain (*Vibrio*). Escape can seldom be directly prevented.

### Transmission

This is how an agent moves from its reservoir to the human host (pathway of exposure). This is clearly the link most amenable to environmental intervention. The routes of transmission include direct exposure to excretions or respirations of humans or animals (difficult to prevent) or indirect exposure through air or vehicles of transmission (food, water, or inanimate objects). In addition, animate vectors (arthropods or other living creatures) play an important role in the transmission of many human diseases.

A vector may have one of two roles. The role of the vector may be that of biological transmission, where it serves as an intermediate or primary host to the agent and is an integral part of the life cycle. Examples include the female anopholene mosquito for transmission of malaria from its human reservoir to other susceptible humans and the snail as an integral part of the life cycle of a blood fluke, the agent of schistosomiasis, also from a human reservoir. Or, the vector may serve only to transfer the pathogen mechanically from the source to a vehicle or to a new host. Houseflies are an example of mechanical transmission when they transport enteric pathogens from manure to a human food item.

### Portal of Entry

Chemical or biological agents gain access to the human body through a variety of mechanisms. The most common of these are inhalation (for respiratory agents), ingestion (for enteric agents), and injection or absorption through the skin (for blood-borne agents). The eye, the ear, and the genito-urinary tract also serve as portals of entry for some agents.

### Host

Host factors play a major role in human health. A healthy immune system provides significant protection, particularly against infectious agents. Immunocompromised individuals become susceptible to conventional pathogens and to the normally innocuous opportunist organisms which will be discussed in Chapter 10. Genetic factors also play a major role in human health. Although host factors are often beyond the reach of environmental intervention strategies, Public Health immunization programs have been enormously important in curtailing many infectious diseases, eliminating smallpox, and nearly eliminating paralytic polio.

## ENVIRONMENTAL INTERVENTION

Prevention of human illness or trauma requires some combination of the following three elements; Technical/Engineering approaches, Legislative/Regulatory approaches, and Education. Technical or engineering intervention can generally be described in the following categories:

### Substitution

Substitution of safer materials or processes has previously been referred to as "prevention." This can also include exclusion or prohibition of known toxins from certain processes; for example, excluding heavy metals or chlorinated plastics from packaging material that will be incinerated.

### Physical Barriers or Enclosures

These can include such diverse approaches as screens to keep out insects, airbags in cars, and lead shielding for X-ray units. This category of intervention includes containment—landfills for solid waste and long-term encapsulation of radioactive wastes are examples.

### Filtration

Filtration is used extensively in water and wastewater treatment processes and for "end-of-pipe" air pollution control as well as in building ventilation systems.

### Ventilation

Using directional air movement and/or dilution minimizes concentrations of contaminants.

### Chemical Neutralization

Examples include raising the pH of acids or reducing the pH of bases.

### Bioremediation

This is an increasingly important tool for remediation of hazardous waste sites or for biofiltration to neutralize organic compounds. New developments in biotechnology are leading to engineering of microbes specifically to digest or convert toxic chemicals in waste sites to innocuous substances.

### Cidal Processes

Cidal processes include microbicides for killing pathogens, rodenticides for killing rats and mice, and insecticides for killing arthropods. Microbicides include wet and dry heat, liquid and gaseous chemicals, and ionizing radiation. The use of chemical pesticides is controversial because useful pesticides are inherently damaging to living cells.

### Biological or Natural Approaches

This concept, which has gained considerable momentum in recent years, includes the introduction of natural predators, the use of synthesized hormones (such as juvenile hormones to control mosquito larvae), and biological pathogens specific for target pests. These and other innovative approaches based on knowledge of the pest will be discussed in Chapter 15.

### The Integrated Approach

Concerns that the use of potent chemicals even for the noblest of public health applications may have unintended deleterious effects on nontarget species (including humans) has fostered the concept of integrated pest management (IPM), which will be discussed more specifically in Chapter 15. IPM starts with detailed knowledge of the epidemiological chain for a given disease, and knowledge of all aspects of the life cycle and habits of the target pest. Then a combination of strategies including technical, chemical, and biological or natural elements is devised specifically for that pest, to minimize unintended effects on nontarget species and the environment.

## SURVEILLANCE OF HUMAN HEALTH PROBLEMS

In addition to a thorough understanding of the epidemiology of a given disease, we want to know something about the impact on the population at risk as measured by morbidity and mortality. Thus, measurements of the incidence, prevalence, and mortality rates need to be obtained. In the United States, the Centers for Disease Control and Prevention (CDC), in Atlanta, Georgia, is charged with that responsibility. Worldwide, the World Health Organization (WHO), headquartered in Geneva, Switzerland, fills that role. However, obtaining timely and complete information on the scope of human illness is no easy task.

In the United States, nationally "reportable" infectious diseases are funneled from physicians making the diagnosis through the state health departments and then to CDC. However, diagnosis is often uncertain or difficult to confirm. Often laboratory identifications are not carried out. Many diseases (such as some food-borne infections) may be sufficiently mild or transient that physicians are not seen. Additionally, there is wide discrepancy among the states in the completeness of reporting. For chronic illness such as cancer the task is even more difficult, although many states have now established cancer registries. The problem is obviously exacerbated in developing nations with inadequate public health infrastructure.

Mortality statistics are somewhat more complete, because death certificates, which list cause of death, are available in every state.

The CDC surveillance system provides valuable information on trends in endemic illness and detects and investigates countless outbreaks every year. This information is disseminated widely through the Morbidity and Mortality Weekly Report (MMWR), available by subscription or on the Internet. An annual summary provides ongoing year-to-year comparisons of incidence and mortality for all of the reportable diseases. Periodic summaries are also prepared on specific acute and chronic illnesses with up-to-date guidelines on prevention and treatment. From these summaries it is possible to rank causes of death and detect trends. Heart disease continues to rank number one and cancer number two on this list.

Another summary is based on years of potential life lost before age 65. This summary recognizes the disproportionate effect of old age (still an incurable affliction) and concentrates on illnesses affecting the very young. In this ranking, unintentional injuries rise to the top, and AIDS, homicide, suicide, and sudden infant death syndrome assume greater importance because they primarily affect the younger segment of the population. Cancer remains in the number two position, emphasizing the importance of that disease among people of all ages. AIDS had reached the number four ranking by 1992 but has since dropped lower as the effects of life-prolonging treatment regimens have taken hold.

The surveillance system becomes all the more important with the increased concern for emerging infectious diseases and the need for early detection of previously unrecognized threats to the population.

For environmental health professionals seeking to prioritize their use of available resources, continued attention to CDC reports is a very necessary exercise. Recognizing the environmental role in causation remains difficult, particularly for chronic illness with its many confounding variables, from genetics to individual variation to incomplete exposure histories.

## REFERENCES

The following are valuable references for anyone interested in current information and trends related to human health and the environment. They have been essential sources of information in my ongoing efforts to stay current on the dynamic changes occurring continually in this field.

Benenson, A. (ed). 1995. *Control of Communicable Diseases Manual.* (16th Ed). American Public Health Association. Washington, DC. (New edition published every five years). Also available on CD Rom.

*Morbidity and Mortality Weekly Report* (MMWR). Centers for Disease Control and Prevention (CDC). Atlanta, GA. (Subscriptions available from Massachusetts Medical Society, P.O. Box 9120, Waltham, MA 02254-9120. (Also available on the Internet using Adobe Acrobat software.)

MMWR, CDC. *Summary of Notifiable Diseases, United States, 1997.* Vol. 46, No. 54, Nov. 20, 1998. (This summary is published annually, usually appearing each fall for the preceding calendar year.)

# Section II-

# THE MACROENVIRONMENT

# Chapter 3

# POTABLE DRINKING WATER

Human civilization has always depended on a reliable supply of safe drinking water for survival, but even though two-thirds of the Earth's surface is covered with water (mostly nonpotable seawater), such a commodity is not universally available. In the United States we have gone to great lengths to provide safe drinking water to the increasing percentage of our population residing in cities and towns. The twentieth century has seen major advances in municipal water treatment accompanied by dramatic decreases in waterborne diseases such as typhoid fever. Nonetheless, several incidents in the 1990s have made us realize that we cannot take these advances for granted.

In 1991, a major outbreak of cholera started in Peru and subsequently spread to some twenty-one countries in the western hemisphere, causing more than one million cases and ten thousand fatalities. Cholera had been totally absent from the western hemisphere from 1911 until the 1970s when a few sporadic cases began cropping up. The recent outbreak was perhaps inevitable given infrastructure problems in Latin American countries where modern wastewater and drinking water treatment plants have not been established. Now entrenched in the human population, and spread by the food-borne and waterborne routes, it will probably remain endemic in the western hemisphere for years to come.

Closer to home, and even more disturbing in its implications for our highly technological society, was the outbreak of cryptosporidiosis in Milwaukee in 1993. Though few people died, this outbreak sickened some 403,000 people. It was traced to a municipal water supply that met all of the potable water standards in effect at the time. An earlier outbreak affecting 13,000 people occurred in Carrolton, Georgia, in 1987.

How could such an outbreak happen in a modern American metropolitan area? The full story may never be known, but a chain of contributing events can be postulated. Cryptosporidium is a protozoan with human and animal reservoirs. In this case, the source may have been dairy cattle. Dairy farms are abundant in Wisconsin, and cow manure is routinely spread on fields in the spring. That year a late thaw may have resulted in greater than normal runoff of the manure into streams feeding Lake Michigan (the source of Milwaukee's water). A combination of tides and offshore winds could have carried higher than normal concentrations of the agent to the inlet for one of two water treatment plants serving the system.

It is known that cryptosporidium is resistant to chlorination, the major mechanism for destroying pathogens in drinking water. Treatment plants are thus dependent on the filtration step for removing the protozoan cysts. Cryptosporidium cysts are relatively small, and thus difficult to filter. Efficient filtration in turn is dependent on a flocculation step which has historically used alum as the flocculation agent. Concerns about aluminum in drinking water have fostered alternate flocculation methods to reduce aluminum content of the finished product. Successful filtration results in very low turbidity. The turbidity level at the time of the outbreak, as measured by light transmittance, met the 1993 standard. However, it is interesting to note that the standard has since been tightened and the 1993 Milwaukee turbidity level would now be above the limit. The infectious dose is also reported to be very low (< 30 oocysts). At any rate, that incident was a wake-up call to municipal water treatment plant operators, and surveillance over cryptosporidium levels has increased dramatically in an effort to prevent recurrences. Attention to the aging process of treatment plants built in the early twentieth century and needs for innovation in treatment have resulted from this incident.

## WATERBORNE DISEASE IN THE UNITED STATES

The number of outbreaks and cases of acute waterborne disease is reported by CDC on a biannual basis. Table 3–1 summarizes data from those reports for the last five reporting periods (there is a lag of three to four years in reporting).

Table 3-1

Waterborne Disease Summary, 1987-1996

| Year | # of Outbreaks | # of Cases |
|---|---|---|
| 1987-88 | 28 | 24,277<br>(13,000 from Carrolton, GA) |
| 1989-90 | 26 | 4,288 |
| 1991-92 | 34 | 17,474 |
| 1993-94 | 30 | 405,366<br>(403,000 from Milwaukee) |
| 1995-96 | 22 | 2,567 |

While the number of reported outbreaks has remained relatively constant over this period, it is clear that the occasional major incident can skew the number of cases. The reports also shed light on the nature of waterborne disease agents. In those outbreaks for which a specific agent can be identified, protozoa are most common, followed by bacteria. Relatively few outbreaks are of chemical origin.

## Chemical Contamination

While our major concern over chemicals in the water supply centers on long-term chronic effects, CDC data are available only for acute poisonings. Chemicals were responsible for only 547 total cases out of the 453,972 reported from 1986-1996. These acute cases involve primarily lead, copper, fluoride, and nitrates. Lead and copper get into water primarily by leaching from pipes under acidic conditions. Drinking water is seldom low enough in pH to cause such leaching, hence the relatively small number of cases. Lead content in plumbing fixtures in contact with potable water must be below 8 percent under the Safe Drinking Water Act. Fluoride incidents have occasionally resulted from accidental release of higher than intended levels during the fluoridation process, which will be discussed later in this chapter.

Nitrates (primarily from well water) remain as perhaps the most significant acute chemical problem in drinking water. High nitrates are responsible for a blood condition called methemoglobinemia in infants less than three months of age. Relatively high pH in the infant stomach permits the conversion of nitrate to nitrite, resulting in an acute cyanotic response. There has also been a possible link to miscarriages reported for high nitrate content of well water. The primary drinking water standard for nitrates has been lowered to 10 ppm. Because nitrates are commonly used in fertilizer, there has been inevitable seepage into wells in agricultural regions. A recent survey found 13 percent of midwestern wells to exceed the primary standard for nitrates.

Much attention has been directed of late to the presence of trihalomethanes in drinking water. These chemicals, typified by chloroform, result from the reaction of chlorine with organic matter. Chloroform is a suspected weak carcinogen in some rodent test animals, suggesting the potential for excess cases of human cancer from consumption over a lifetime. The standard in drinking water has been set at 50 ppb. While most municipalities can meet that standard relatively easily, the concern has led to changes in chlorine application to minimize trihalomethane production.

## Infectious Agents in Drinking Water

Historically, it has not always been possible to identify specific infectious agents in outbreaks epidemiologically linked to drinking water. Historically, some 50 percent of outbreaks have been reported as acute gastrointestinal illness (AGI) without specific etiology. It is theorized that most of those outbreaks were caused by viruses (such as Norwalk virus) which have been difficult to identify. Improved identification techniques should help to reduce that number of unsolved outbreaks. In the last reporting period, agents were identified in 63.6 percent of the outbreaks, indicating progress in rapid identification methods.

For those outbreaks in which an agent is identified, protozoan etiology is by far the most common. The importance of cryptosporidium has already been discussed. The other commonly reported protozoan agent is *Giardia lamblia*. Giardia is actually the most commonly reported water pathogen in terms of number of outbreaks. It is mostly of human origin although a reservoir in beavers is also suspected. Common among campers obtaining drinking water from surface sources such as streams, giardia has also been commonly reported among international travelers, particularly those returning from Russia. A third protozoan disease is amoebiasis, although many fewer clinical cases have been reported in recent years.

Viral agents important in drinking water include hepatitis A and E, Norwalk virus, and rotavirus. All enteric viruses are susceptible to conventional treatment. Rotavirus was first identified in 1973 and is sometimes considered the first of the emerging infectious diseases. It affects only infants and children, with adults being resistant except to extremely high doses. Hepatitis A and E and Norwalk virus will be discussed further in Chapter 12 as food-borne agents.

Among the bacterial agents which have caused waterborne illness, the most prominent have been *Salmonella* spp. (including typhoid fever), *Shigella, Campylobacter, Vibrio* spp. (including cholera), and *Yersinia. Escherichia coli* 0157:H7 was also implicated in numerous recreational water outbreaks in the latest (1995-96) reporting period.

## Waterborne Disease and Recreational Water

In addition to disease transmission through ingestion of contaminated drinking water, numerous outbreaks have been associated with bathing in recreational water. The same pathogens implicated in drinking water outbreaks are obviously transmissible in swimming pools and beaches, as bathers often ingest large quantities of water. The important food-borne pathogen *E. coli* 0157:H7 was also implicated in a recreational water outbreak for the first time in 1991 and was responsible for six outbreaks in the 1995-1996 reporting period. The largest number of cases in 1995-1996 was attributed to *Cryptosporidium parvum. The* potential for direct release of enteric pathogens from bathers, particularly small children, is high, and even in chlorinated recirculating swimming pools complete destruction of pathogens cannot be assured. Eye and ear infections are additional threats in recreational water with those portals of entry being exposed to contaminated water. Finally, there have been numerous outbreaks of dermatitis folliculitis (a skin condition) caused by *Pseudomonas* bacteria, associated with recreational hot tubs and whirlpools. Such infections can be prevented only through rigid attention to cleaning and disinfecting the facilities. Even then, the communal bathing use of recreational water makes it difficult to assure that such problems will not occur.

### Aerosolization from Water Reservoirs

Legionellosis is a unique respiratory disease caused by a bacterium (*Legionella pneumophila*) aerosolized from a warm water reservoir. This disease will be discussed in Chapter 5, on air pollution.

## DRINKING WATER REGULATIONS

In the United States, potable drinking water is regulated by the Safe Drinking Water Act (Public Law 93-5-23), which is enforced by the Environmental Protection Agency (EPA). The Act went into effect in 1978 and has been reauthorized and amended numerous times since. Before this Act there were no national standards for the safety of potable water. Federal standards applied only to interstate commerce (trains, airplanes, ships) but many states adapted those standards for their own use. The safe drinking water standards are themselves somewhat controversial and difficult to enforce. While 1986 amendments required that the EPA set standards for numerous new chemicals on an ongoing basis, money has not been forthcoming for enforcement of such an ambitious undertaking. In 1994, it was estimated that 8 percent of U.S. public water supplies exceeded maximum contaminant levels (mostly the microbiological standard). The 1996 reauthorization required public notification when standards were exceeded. The standards are divided into primary (health-related) and secondary (non-health-related ) sections.

### Inorganic Chemicals

The primary standards include many inorganic chemicals, with limits set in mg/L. For example, Barium (1.0), Chromium (0.05), Cadmium (0.01), Lead (0.05), Mercury (0.002), Nitrates (10), Selenium (0.01), and Silver (0.05) are on the list. Aluminum is an interesting example of an inorganic chemical slated for a standard. Used extensively in water treatment as a flocculating agent, aluminum has been linked circumstantially to Alzheimer's disease and also to the leaching of lead from water pipes. However, hard evidence of its effect on human health when ingested in the water has not been forthcoming.

### Organic Chemicals

Organic chemicals have not been linked directly to acute toxicity when ingested in drinking water. I have previously discussed the trihalomethanes associated with chlorination and the standard of 50 ppb set for chloroform based on the tenuous extrapolation to excess cancer cases. Volatile organic compounds (VOCs) must also be monitored under the latest standards. Polychlorinated biphenyls (PCBs) are an example of a persistent water contaminant which does not reach sufficient levels in drinking water (upper limits of contamination are about 3 ppb) to cause concern.

However, its role in the food chain through bioaccumulation in fish has been associated with fish advisories, which will be discussed in Chapter 13.

### Microbiological Standards

Infectious agents are the major acute concern for safety of the drinking water supply. Unlike chemical content, the microbiological makeup of the supply is subject to frequent change. Because it would be impractical to test for the array of potential pathogens on an ongoing basis, an indicator organism is used as the basis for determining microbiological safety.

For many years that indicator has been the fecal coliform group of bacteria. They are natural inhabitants of the intestinal tract of humans and other warm-blooded animals, and thus serve as a surrogate for pathogens associated with fecal contamination. Although they are less indicative of viral contamination or of protozoan cysts, they serve well as indicators of water quality. They are relatively easy to test for; they are very likely to be present when other intestinal pathogens are present; they are as resistant to chlorination as other enteric bacteria or viruses; and they are mostly nonpathogenic themselves to healthy humans. While other indicators have been proposed, none has demonstrated any real advantage over the coliforms as a general measure of potability. The standard is <1/100 mL of water tested, with an action level set at 4/100 mL. No more than 5 percent of samples may be positive on a monthly basis. When positive samples are found, the sampling is repeated, and steps (such as increasing the chlorine level) must be taken to maintain the negative coliform status of the supply.

### Other Primary Standards

Other primary standards include radionucleide levels and turbidity. Turbidity was once considered secondary but is now primary because excessive turbidity interferes with effective disinfection and filtration. As mentioned earlier, the turbidity standard has been strengthened to counter the growing problem with cryptosporidia.

### Secondary Standards

These standards govern factors such as color, odor, dissolved solids, hardness, and acidity. Some examples (in mg/L) include chlorides (250), sulphates (250), iron (4) and manganese (0.05). Water hardness has posed an especially intriguing question for environmental health professionals. Hard water (caused by calcium and magnesium ions) interferes with detergency, and many municipal supplies and private wells are equipped with softeners to counter that problem. However, there is considerable evidence that magnesium, in particular, provides protection against ischemic heart disease, raising the issue of the overall benefit versus risk of deliberately lowering the magnesium level by water softening. As is often the case, direct epidemiological evidence of the protective effect of magnesium in drinking water is unavailable, as magnesium is also obtained by dietary means.

### Recreational Water Standards

Coliform bacteria have also been used as the indicator organism for recreational water, although again with some controversy as to whether or not they are the best indicator available. The standard has been set at 200 coliform/100 mL. In some northern states (Wisconsin and Minnesota), the level is only enforced seasonally from March to November under the theory that wastewater need not be chlorinated during months when there is no recreational use of the receiving stream. Thus, trihalomethane levels can be concurrently reduced without compromising human health.

## WATER SOURCES, TREATMENT, AND USE

Domestic water use in the United States requires approximately 150 gallons per capita per day. Only a small fraction of that quantity is for drinking or culinary purposes. Most is used for bathing, washing dishes and laundry, flushing toilets, watering lawns, and washing cars. For practical purposes, however, it must all be delivered to the consumer as potable water meeting the primary and secondary standards described above. Thus, an economically viable system must be available to provide a dependable supply on a continuous basis. Fresh water is available as a recycled commodity involving precipitation, runoff into surface streams, rivers, and lakes, or seepage into the ground, where it becomes part of the groundwater system. Ultimately, the supply is tapped for domestic, commercial, or agricultural use or simply evaporates to reinitiate the cycle. In the U.S. approximately 50 percent of consumers utilize groundwater while 50 percent are served by surface water systems.

### Groundwater

Groundwater is recovered primarily through the mechanism of drilled wells (private or municipal) tapping into suitable aquifers at varying depths. Such water is often of excellent quality, having been naturally filtered for removal of particulate contaminants including pathogens. Chemical content may be highly variable and sometimes requires treatment for removal of nuisance chemicals such as iron, which can stain porcelain sinks or clothes being laundered. Groundwater is often high in magnesium and calcium, requiring softening, which has been previously discussed. Softening can be accomplished by several mechanisms, typically using sodium zeolite, lime, or soda ash to convert the minerals to insoluble compounds which can then be filtered. Many municipalities also routinely chlorinate wells for extra assurance of a coliform-free supply. An alternative mechanism for utilizing groundwater is the development of springs through which groundwater surfaces naturally. Springs must be protected by housings to prevent surface water from entering.

## Surface Water

Large cities usually must develop surface water systems to assure consistent delivery of the large quantities needed. Many of these cities take the water directly from rivers or lakes while others (such as New York City) utilize more remote watersheds to feed surface reservoirs. Elaborate regulations are enforced to protect the watersheds from gross industrial or domestic wastewater pollution. However, all surface water is subject to natural pollution from birds, animals, and natural runoff of stormwater.

## Treatment

While every municipality tailors its treatment system to particular needs and source water quality, a generic sequence of treatment steps can be described.

### *Natural sedimentation*

Systems that utilize watersheds to fill man-made reservoirs depend on natural sedimentation to improve water quality. Water is allowed to remain in these reservoirs while heavier solid particles settle out, thus providing a natural "treatment" process.

### *Aeration*

Using a fountain effect to aerate the water is commonly used as a means of promoting oxidation reactions, primarily for improvement of taste and odor.

### *Flocculation/Coagulation*

Alum has traditionally been used to induce aggregation of particulates and to precipitate minerals, which can then be removed in the filtration process. As mentioned earlier, concerns about aluminum levels in the water have led to a search for an alternative to alum for flocculation enhancement. Polyaluminum hydroxychloride is one such proposed substitute.

*Filtration*

During the twentieth century, rapid (or pressure) sand filtration gradually replaced the original slow sand filters which were first used for filtering water. Requirements for greater quantities to supply growing metropolitan areas fueled the change, and improved flocculation systems enabled the new filters to match the high percentage removal (~99%) of the slow sand variety. The emergence of protozoan pathogens has placed increasing importance on the maintenance and operation of this treatment step. Lapses in backwashing or flocculation may have contributed to the cryptosporidium outbreaks discussed earlier.

*Disinfection*

I have discussed at some length the importance of chlorination to the provision of safe water. A case can be made that water supply chlorination has been the single most important environmental health intervention of the twentieth century. Chlorine is relatively inexpensive, easy to regulate and inject, and can easily be tested to measure residual immediately following treatment and throughout the distribution system.

While chlorine chemistry is complex, the disinfection process has been well defined. Chlorine is a very reactive halogen. The amount added is based on chlorine demand which is a measure of the organic load which must be overcome. The amount added must be sufficient to react with the organic compounds, kill any pathogens present and leave a measurable residual as evidence that it has done its job. Two types of chlorine are involved. Free available chlorine (hypochlorous acid) is the reactive form, which is readily dissipated as it works. Combined chlorine (chloramine) is less reactive (and therefore less effective as a disinfectant) but persists for long periods, thus providing a residual (typically 1-3 ppm) which can be measured at any point up to the end of the distribution system.

Ozone ($O_3$) has been suggested as a possible substitute for chlorine to mitigate the trihalomethane problem. Ozone is also more effective against cryptosporidium cysts than chlorine. It is used extensively in Europe for water treatment. However, ozone is more expensive and leaves no residual to confirm the process. A recent trend in the U.S. has been to combine the two disinfectants for optimum effect. Ozone is used early in the treatment process, then chlorine is added later to provide residual in the mains. Some 200 cities in the United States were using some ozone in the treatment process as of 1998.

*Fluoridation*

The fluoridation of drinking water has been another major public health intervention in the twentieth century. It has been demonstrated to reduce dental caries in children by more than 65 percent. Unfortunately, fluoridation has not been accepted universally, by all communities, and as the twentieth century ends, only about 60 percent of public supplies are using this proven effective means of improving dental health. Like chlorine, fluoride can be added safely and economically at a level of approximately 1 ppm to provide protection against dental caries. Opposition has centered around the concern that fluorides are carcinogenic (although there is no credible evidence that it is). Fluorides are known to be toxic at high levels. There is a primary drinking water standard for fluoride at 4 ppm, and public notice must be provided if the level exceeds 2 ppm. However, regulation of the amount added is not difficult and incidents of overuse have been few and far between.

Recent comparisons of communities that provide fluoridation with those that do not have shown less of a protective effect on childhood dental caries than previous studies. The reason probably relates to availability of alternative fluoride sources, particularly in toothpastes and direct dental treatments. However, it can also be argued that such alternative sources are less available to lower socioeconomic groups. Thus, the "net societal benefit" concept still supports the municipal fluoridation practice.

*Other treatment processes*

For particular municipal (or private) facilities, additional treatment may be needed to prevent or alleviate problems. Some examples include copper sulfate for algae problems, which typically arise in late summer in some lakes and streams; activated carbon for adsorption of synthetic organic compounds (SOCs); packed tower aeration for removal of volatile organic compounds (VOCs); and sulfur dioxide to dechlorinate in systems where high chlorine amounts are added and there is need to reduce the potential for trihalomethane production.

## Sea Water

In some parts of the world, freshwater is simply not available in quantities necessary to support the population. One option available to some is the utilization of sea water following desalinization. The salt can be removed through a distillation process. However, distillation is expensive and energy intensive. Thus, the process is used only in areas, such as the Middle East, where freshwater is particularly scarce and energy is relatively abundant and affordable.

## Reuse of Water

Another option that is now considered for water-short areas is the reuse of wastewater. The concept is that wastewater treatment, which will be discussed in Chapter 4, can be sufficiently effective to provide a viable source of water to be further treated for public distribution in the potable state.

**REFERENCES**

Centers for Disease Control. 1998. "Surveillance for Waterborne Disease Outbreaks- United States, 1995-1996. *Morbidity and Mortality Weekly Report.* Vol. 47 No. SS-5. Pages 1-33.

Eaton, A. D, L. S. Clesceri, and A. E. Greenburg. 1995. *Standard Methods for the Examination of Water and Wastewater.* (19th ed). American Public Health Association, American Water Works Association and Water Pollution Control Federation. Washington, DC.

Salvato, J. 1992. *Environmental Engineering and Sanitation.* (4th Ed.). Wiley-Interscience. New York. Chapters 3 & 4.

# Chapter 4

# WASTEWATER

Once potable water has been distributed through the mains to the consumer, it serves as a carrier of multiple pollutants added through domestic, commercial, and industrial use. Domestic (or sanitary) sewage—the wastewater discharged by household and commercial users—has the obvious high potential for carriage of pathogens from human wastes, dishwashers, laundry, bathing facilities, and miscellaneous substances that get flushed down toilets or drained into sinks.

Industrial wastewater has the added potential for chemical or biological loading from the particular process being carried out. Some industrial facilities, such as dairy plants or paper pulping operations, produce waste with high organic content, which should be reduced before discharge. Other industries may discharge a variety of hazardous chemicals. Discharge of such wastes often requires a permit, which will specify how the waste must be treated before it is released into the public sewer. A rule of thumb is to require treated water to equal the quality of that provided by the community to the industry.

In some areas, the sewage system also handles storm runoff from public streets. At times of heavy rainfall, that added volume can overwhelm the treatment plant, resulting in bypass directly to receiving streams. In many metropolitan areas storm sewers have been entirely separated from sanitary sewers to preclude that problem.

Wastewater collected through sanitary, industrial, or storm sewers and ultimately discharged (treated or untreated) into a receiving body of water is referred to as "point source" pollution. Such sources can be monitored to determine the strength of pollution. "Nonpoint" sources include agricultural fields (with high nitrate and phosphate content), feedlot operations (with high coliform counts), residences, parks, and golf courses (which use lawn chemicals). Because such pollutants are not collected in identifiable sewer systems, the discharge is more difficult to monitor. Natural runoff from fallen leaves, feral animals, and birds is also in this category, adding organic content to the water.

## POLLUTION MEASUREMENTS

The primary measurements of water pollution are biochemical oxygen demand, dissolved oxygen, and total suspended solids. Biochemical oxygen demand (BOD) is defined as the amount of oxygen needed to stabilize the organic content during a five day test. Dissolved oxygen (DO) measures the amount of oxygen available to cope with the BOD load. Total suspended solids (TSS) is a measurement defining the turbidity of the wastewater.

## Stream Pollution and Recovery

Let's consider, by way of example, an initially clean river or stream. There will be an excess of DO compared to BOD. If a point source discharge of high-BOD wastewater occurs, the stream will enter what is called a "zone of degradation." As the DO attempts to stabilize the BOD, both measurements will decrease. If the oxygen is exhausted before the BOD is satisfied, the stream enters a "zone of active decomposition." In this situation only anaerobic life forms (some helminthic forms and anaerobic bacteria) survive. Fish and green plants are totally absent. If no further pollution is added, the stream will gradually enter a "zone of recovery" as natural turbulence adds oxygen and the BOD is further reduced. Eventually, the stream again enters the clean water phase with an excess of oxygen over BOD. This process is illustrated in Figure 4-1.

Figure 4-1

Stream Pollution and Recovery

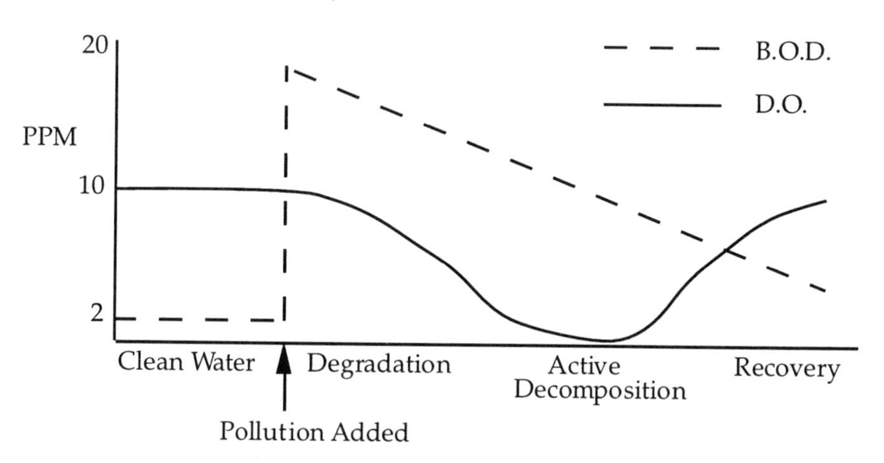


A similar situation marks the phenomenon of eutrophication often applied to lakes. Eutrophication occurs when runoff of phosphates and nitrates from agricultural or human sources provides an excess of nutrients to green algae in the lake. The algae proliferate to extremely high concentrations, then begin to decay. The decay process depletes the oxygen and the lake goes to extinction. This situation occurred in Lake Erie in the 1960s, garnering considerable publicity. Commercial fishing and recreational use were greatly curtailed. However, once more stringent pollution control regulations were imposed and enforced, the lake demonstrated a strong comeback, with normal use being restored by the 1980s.

### Thermal Pollution

One other form of water pollution that has received some publicity is thermal pollution. This occurs principally from hot water discharges from power plants located on rivers or lakes. The elevated temperature can adversely affect many wildlife species. However, in northern climates, it can also prevent freezing, leaving open water necessary for many species of waterfowl.

## HUMAN HEALTH IMPLICATIONS OF WASTEWATER

In the United States, the major direct human health implication of polluted water is contamination of recreational water for swimming. There is also the potential for discharge from septic tank systems to contaminate drinking water wells and the obvious implication of more difficult downstream water treatment. A river such as the Mississippi provides drinking water for numerous cities along its banks from the Twin Cities in Minnesota to New Orleans in Louisiana. Those cities also discharge their wastewater back into the river for use downstream. (This can be viewed as a water reuse example as described in the previous chapter.) One additional human health implication is the occupational exposure of wastewater treatment plant workers to aerosols produced during treatment, although deleterious health effects have not been definitively demonstrated.

In developing countries, there is a shortage of wastewater treatment facilities, leading to serious disease transmission problems. Schistosomiasis, which will be discussed in Chapter 14, is a major debilitating disease affecting millions of people in endemic third world countries. It is a prime example of a disease that cannot be prevented without infrastructure changes to provide modern treatment and disposal of human wastes. Cholera, which was discussed in Chapter 3, is another example.

Finally, there is the largely unknown effect of the many trace chemicals that contaminate bodies of water, sometimes many miles from the point of origin. Industrial discharges such as mercury can be carried by air currents and deposited in bodies of water hundreds of miles from the source. Persistent chemicals such as PCBs and DDT appear in water and concentrate in fish many years after a ban on their use. Pesticides such as Toxophene appear in water samples with no apparent source. Deformed frogs are numerous in some locations, and chemicals in the water are a suspected but unproven cause.

## WASTEWATER TREATMENT

From a regulatory perspective, wastewater discharge standards have been covered by the Clean Water Act first passed in 1948. The Act covers only point sources, although, as mentioned earlier, nonpoint sources are currently receiving more attention. Current requirements were primarily set through amendments passed in 1987.

### Individual Wastewater Systems

While the infrastructure of municipal sewer lines in the United States has been extended widely during the twentieth century, many rural areas are still dependent on individual systems. Dwellings without running water must, of course, depend on outhouses with their inherent problems of odor, insects, and potential overflow of untreated human waste.

In rural settings with individual wells, water carriage systems provide the opportunity for individual septic tank systems. In less densely populated areas, properly maintained septic tanks can provide a satisfactory means of wastewater management. A septic tank is a watertight chamber designed to hold at least one day's wastewater discharge from a dwelling or building. Baffles assure adequate holding time for the wastewater. Anaerobic bacteria serve to digest and liquify solids.

The liquid effluent is then discharged through a distribution box to a series of perforated tile pipes laid in trenches of varying length. The number and length of drain pipes is calculated based on the expected daily flow and the porosity of the soil. Soil porosity is determined by a percolation test to determine how rapidly the water will drain. Gravelly or sandy soil drains best while heavy clay soils drain poorly and may be unsuitable for this type of system. Ideally, the drain field should be in an area sufficiently above bedrock and above the expected high water table level to allow adequate filtration to occur.

An alternative to the drain field is a seepage pit, which may be a less expensive choice in an area with porous soil and deep bedrock and water table. Not surprisingly, such systems are frequently installed under less than ideal conditions and failures are common. Such failures are usually marked by pooling of the effluent (which is still subject to high coliform counts) on the ground surface, causing odor problems, attracting vermin, and potentially exposing children and pets to dangerous pathogens. Even the best systems must be periodically pumped out to dispose of solid materials not digested anaerobically.

In the years following World War II, rapid expansion of suburbia often outpaced the expansion of municipal sewers, and large subdivisions sometimes depended on individual systems. The combination of multiple septic tanks and individual wells was the source of frequent cross-contamination, with wells becoming contaminated from septic tank effluents, particularly when soil conditions did not permit adequate drainage. Fortunately, such situations are rare now.

## Municipal Wastewater Treatment

Wastewater treatment facilities are intended to convert the noxious discharges of residences, commerce, and industry into clean water, posing no threat to health or recreation. As I have previously discussed, discharged wastewater frequently becomes the drinking water source for downstream communities. Although wastewater treatment plants are individually designed, a generic description can outline the elements of treatment most widely practiced in the United States currently.

### *Mechanical treatments*

Even assuming that storm sewers have been separated from the sanitary sewers, there will be some solid debris (sticks, stone, glass, etc.) in the influent wastewater. A bar screen or comminutor (a type of solid waste grinder) is frequently used to screen out this solid material. This is often followed by a "grit chamber," or widening of the channel designed to reduce the velocity of the stream to allow heavier particulates (sand and gravel) to settle out.

### *Primary wastewater treatment*

The next step is frequently a primary settling tank, designed to hold the wastewater for a sufficient time to allow further settling of solids. Flocculants may be added at this point to abet the process. The solid material from the primary settling tank is referred to as "sludge" and is transferred to a primary digestor, which functions much like a septic tank in an individual system. The sludge digestor is based on anaerobic digestion of the solids. It may be seeded with anaerobic bacteria. The temperature is controlled and the sludge is agitated to optimize the anaerobic activity. Once digested, the sludge is removed to drying beds, where a relatively dry "sludge cake" is formed. The sludge becomes a solid waste, which can then be buried, burned, or utilized as a fertilizer to be spread on agricultural land.

This recycling process now accounts for about 50 percent of the sludge disposal total, although not without controversy. Opponents worry about possible contamination with heavy metals or pathogens. The weight of scientific evidence appears to validate the process as a safe and responsible means of recycling. From an environmental health perspective, heavy metals should be kept out of the wastewater stream through responsible commercial and industrial practices and through education of householders. Pathogens are destroyed by the temperatures reached in sludge digestion and are not a credible threat in sludge used as fertilizer.

*Secondary wastewater treatment*

While some smaller cities and towns may still use only primary treatment before discharging wastewater, most municipalities now employ at least secondary treatment as well. Secondary treatment utilizes aerobic bacteria and other life forms to promote oxidative conversion of sewage solids from noxious to more benign compounds. It is carried out in one of two principal systems.

In the *trickling rock filters* system, a rotating arm sprays the sewage (the effluent from the primary settling tank) over a circular bed of crushed gravel. The gravel has been seeded with a variety of aerobic bacteria, protozoa, and helminths. These form a "zoogleal mass," which eats the aerated spray as it passes through the crushed gravel filter.

In the *activated sludge chamber* system, a similar array of aerobic life forms are seeded in a tank. The effluent from the primary settling tank is pumped into the activated sludge tank, incorporating oxygen during the pumping process.

The effluent from the secondary treatment process flows into a secondary settling tank where any remaining undigested solids are allowed to settle out. Those solids are recirculated back through the anaerobic sludge digestor.

*Tertiary wastewater processes*

Many municipalities now incorporate additional tertiary processes into wastewater treatment. These treatments are based on specific needs to remove additional soluble contaminants not affected by primary or secondary treatment. Examples of tertiary treatments include the following:

1) Denitrification, 2) Ammonia stripping, 3) Chemical precipitation, 4) Carbon adsorption, 5) Electrodialysis, 6) Ion exchange, 7) Distillation, 8) Freezing, 9) Reverse osmosis, 10) Bioremediation, and 11) Biofiltration.

*Disinfection*

There is some controversy over the value of chlorinating wastewater effluent before returning it to receiving waters. The previously described processes do not selectively eliminate pathogens from the effluent, and high coliform counts are inevitable in those effluents. Coliform survival in relatively clean streams is less certain. Generally, chlorination of the effluents is practiced to protect recreational users of the water. The major reason for not chlorinating is to reduce levels of trihalomethane, particularly in water to be used for drinking by downstream communities. As mentioned in Chapter 3, in some northern states, chlorination is practiced only during recreational-use months, typically March to November. When chlorination is practiced, the effluent must be held for a sufficient time to allow the disinfectant reaction. The holding period could also provide the opportunity for dechlorination if desired. A generic municipal wastewater process is diagrammed in Figure 4-2.

Figure 4-2

Municipal Wastewater Treatment Scheme

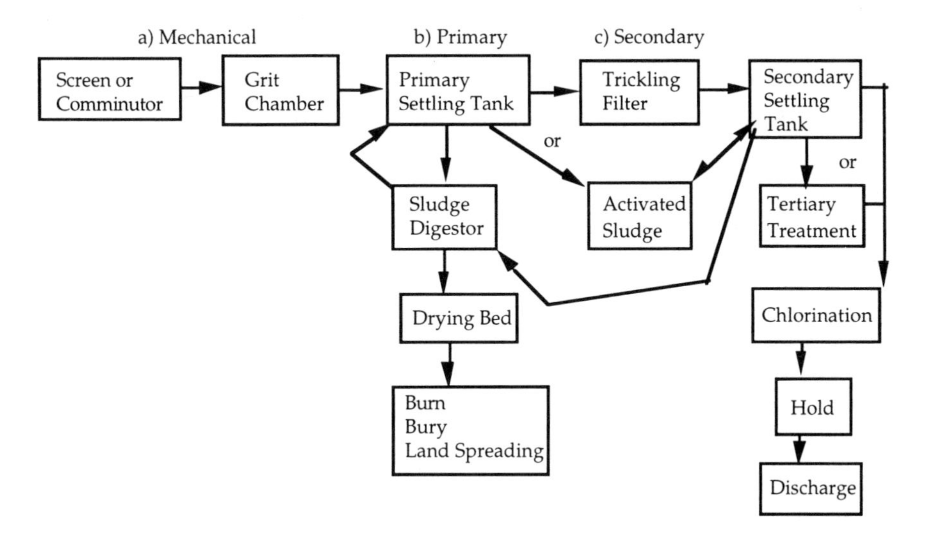


*Alternative wastewater treatment*

In some parts of the United States (such as California), climate conditions are suitable for a less expensive wastewater treatment process utilizing natural processes in a stabilization pond (or lagoon). Typically about six feet deep, and 90 by 300 feet, these ponds provide both anaerobic and aerobic action. By controlling the flow rate through the pond, aerobic activity near the surface and anaerobic activity near the bottom can proceed simultaneously. If properly managed, aesthetic concerns with odor and vermin can be minimized.

# Chapter 5

# AIR POLLUTION

Ambient community air pollution has long been at the forefront of environmental health concerns. It is high on the list of "outrage factor" complaints because the right to breathe healthy air is considered sacred, and the public discerns little individual control over air quality. Air pollution is defined as foreign substances in the air which have an adverse effect on human health, animals or plants, or are damaging to property. As we shall see, the association between air pollution and human health is often unclear for the same reasons that the relationship between the environment and human health in general is unclear, as discussed in Chapter 1: few direct, acute effects and uncertain long-term, chronic effects.

Air is composed naturally of oxygen (21%), nitrogen (78%), carbon dioxide (0.03%), and water vapor (which varies with relative humidity). Carbon dioxide is the primary greenhouse gas being monitored as an indicator of global warming. It has indeed increased since the industrial revolution, reaching a 160,000-year high of 364 ppm in 1997 according to the Worldwatch Institute. While the effect of that rise on global warming is still debated, it has no direct effect on human health. In fact a level of 1,000 ppm is often used as an indicator of indoor air quality problems. As progress has been made in reducing ambient community air pollution levels, much of the attention to air pollution has shifted to indoor air quality in the later years of the twentieth century.

Since the passage of the national Clean Air Act in 1970, and promulgation of standards under that Act, ambient air quality has measurably improved in many regions of the United States. Comparing the major sources of air pollution as measured in 1970 to those in 1992 gives some clues as to changing patterns of air pollution. Table 5-1 shows composite estimates for the United States, obviously subject to wide local variation. The variation is greater worldwide. Volcanic eruptions, for example, have substantially increased atmospheric particulates, sometimes for many years. Major meteor impacts have had even greater influence on long-range particulate concentrations and are blamed by many scientists for the abrupt extinction of the dinosaurs (and 70 percent of all species) 65 million years ago. (Particulate pollution was so dense that sunlight couldn't penetrate, which is the same effect caused by nuclear winter. Extinctions resulted from lack of food after extremely low temperatures and light limited plant growth.) In 1997, extensive deliberate burning of forests in Southeast Asia and Indonesia, combined with delayed monsoon rains, contributed to serious particulate air pollution throughout that region.

Table 5-1

Sources of Outdoor Air Pollution (Composite U.S. %)

| Source | 1970 (historic) | 1992 (EPA) |
|---|---|---|
| Transportation | 60 | 56 |
| Heat and power | 20 | 24 |
| Industry | 17 | 9 |
| Refuse burning | 3 | 8 |
| Community/natural | <1 | 3 |

Although the changes may not seem extensive, they probably reflect improvements in vehicle exhaust control, a crackdown on industrial emissions, and increased utilization of municipal waste incinerators.

## CRITERIA POLLUTANTS

In the Clean Air Act of 1970 (and subsequent reauthorizations and amendments), the emphasis has been on criteria pollutants, thought to be the major contributors to the problem. Those criteria pollutants include:

### Particulates

Now regulated in terms of $PM_{10}$ (nonspecific particulate matter $<10$ μm diameter), this standard has become controversial. There are estimates that these particles, arising from multiple mechanical grinding, combustion, and erosion activities, cause more than 60,000 fatalities annually in the United States. There is new emphasis on even smaller particles ($<2.5$ μm), which arise from combustion of coal, diesel, and other fuels. Industry claims, however, that the cost of compliance is not warranted by known health effects.

### Gasses

The gaseous criteria pollutants include:

*Sulfur dioxide* ($SO_2$), which is the major cause of acid rain and the suspected cause of respiratory ailments associated with acute air pollution incidents to be discussed.

$SO_2$ gets into the air principally from coal-burning electric power generating plants, but with contributions from numerous other sources.

*Nitrogen dioxide* ($NO_2$), which comes largely from vehicle exhaust and power plants. $NO_2$ is a major ingredient of smog (the brown haze which results from sunlight reacting with pollutants such as $NO_2$), which afflicts so many cities. $NO_2$ also causes respiratory distress at high levels.

*Carbon monoxide* (*CO*), which is a product of incomplete combustion. Unlike the other criteria pollutants, CO is definitively associated with acute health symptoms and is the cause of approximately 1,000 fatalities every year in the United States. Most of the fatalities occur indoors resulting from faulty furnaces or other gas appliances. Others occur in closed motor vehicles with leaky exhaust systems or tailpipes blocked by snow. New housing codes require that significant direct fresh air be supplied to furnaces. In addition, CO detectors are highly recommended for installation in residences. Ambient outdoor levels seldom reach concentrations high enough to cause acute health effects.

*Ozone* ($O_3$) is included as a criteria pollutant because it has also been associated with acute respiratory symptoms and is believed to aggravate asthmatic attacks. It is suspected of contributing to hypothermia deaths in the elderly. Ozone is generated naturally by sunlight acting on oxygen in the atmosphere and by lightning. It is also an unintended byproduct of electrical fields and is produced from welding equipment, copying machines, and ultraviolet light sources. Ozone in the stratosphere protects the Earth against ultraviolet rays. It also can be used as a disinfectant. Thus, ideally, we would have more ozone at higher elevations while minimizing the concentration in the human breathing zone.

## HUMAN HEALTH EFFECTS OF AIR POLLUTION

### Acute Incidents

Long-term effects of trace contaminants in ambient air remain speculative due to multiple confounding factors such as occupational exposures and tobacco use. However, numerous acute air pollution incidents have yielded insights into the effects of polluted air on human health. Perhaps the most widely studied such incident occurred in the industrial town of Donora, Pennsylvania, in 1948. A similar incident was reported in the Meuse Valley, in Belgium, in 1931 but was not as intensively investigated. Both incidents had the classic elements which we now realize as potentially dangerous to human health, including death, particularly among the elderly and people with preexisting respiratory conditions. Both incidents occurred at a time before there was regulation of emissions from potential air pollution sources. There were no monitoring programs in effect, thus we can only speculate as to the specific concentrations of any given pollutants. The elements that were in place and are now recognized as contributing factors include the following:

*Persistent weather inversion*

Atmospheric inversions are very common in most cities of the world. They occur when air temperatures reverse the normal pattern of being warmer near the ground and colder at higher elevations. Under those conditions there is approximately a 5.5° F decrease in temperature for every 1,000-foot elevation in altitude. Because warm air tends to rise, and colder air tends to sink, there is a natural mixing of the air layers, abetted also by normal wind conditions. Inversions occur, particularly in spring and fall when ground temperatures are warming and cooling, respectively. During an inversion, the air is cooler near the ground and warmer above, and there is minimal mixing. The effect is like a lid that prevents contaminants from escaping. Most often the inversions are transient and normal conditions return within hours. Sometimes, the inversion remains stable with minimal wind for a prolonged period. In the Donora and Meuse Valley incidents the inversions lasted at least five days. The inversion concept is depicted in Figure 5-1.

Figure 5-1

The Inversion Phenomenon

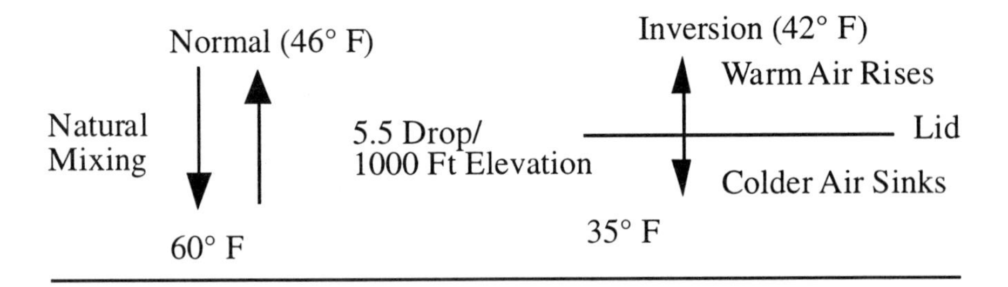


*River valley topography*

Both cities are located in narrow river valleys where high hills surrounding the towns served to restrict horizontal dilution while the inversion was restricting vertical dilution. The effect was akin to placing a lid on a bowl in which pollutants continued to concentrate for the duration of the inversion.

*Heavy concentration of industry*

Both areas were heavily industrial with an assortment of factories, smelters, and refineries operating full blast during the inversion period. Again, because neither regulations nor monitoring stations were in place, it can only be speculated as to ambient concentrations of given pollutants.

In the Donora incident, more than 50 percent of the city's 12,000 residents were made ill. Later incidents in large metropolitan areas served to call further attention to the hazard potential of air pollution. The London area in 1952 and New York City in 1966 both experienced prolonged inversions, both in the autumn and at times when emission restrictions were minimal. Death certificate records indicated several thousand deaths associated with those conditions. Subsequent clean air regulations have averted recurrences in those cities in recent years.

A number of cities in the United States continue to suffer from chronic inversion and smog problems. The most publicized of these has been the Los Angeles smog problem of the 1960s and beyond. The Los Angeles basin is surrounded by mountains on three sides that trap pollutants when winds blow from the ocean. Inversions are frequent although usually transient. Heavy reliance on the automobile and proliferation of a freeway system spawning massive traffic jams contribute to the problem. Various stages of "smog alert" enable public officials to restrict industry, curtail traffic, and cancel outdoor recreational activities when inversions persist and pollution levels (which are now being carefully monitored) rise.

The infamous smog often restricted views of the surrounding mountains during the peak of the problem during the 1960s. As a result of this condition California enacted the most restrictive air pollution emission standards in the country, imposing standards stricter than the national Clean Air Act on vehicles, industry, and public utilities. Solid waste incineration was restricted completely. The result has been a dramatic reduction of the Los Angeles basin smog. However, frequent inversions still occur and routinely produce smog sufficient to restrict mountain views. Other western cities, notably Denver and Phoenix, also frequently experience noticeable smog conditions, eliciting strict regulations and prohibitions similar to those enacted in Los Angeles.

Other cities also experience major air pollution problems. Mexico City is sometimes considered the most polluted metropolitan area in the world. That city is also surrounded by mountains. A burgeoning population (approaching 25 million), increasing industrialization and vehicle traffic, and a lagging regulatory structure have all contributed to the problem. In addition to previously mentioned pollutants, the city is plagued by high coliform and endotoxin levels in the air as a result of inadequate wastewater treatment facilities. Coliform bacteria (and resulting endotoxins) are probably aerosolized from running streams of wastewater in open ditches.

In the summer of 1997, the city of Paris experienced prolonged inversion conditions which forced imposition of traffic restrictions. The previously mentioned situation in Indonesia and parts of Southeast Asia in 1997 is an additional reminder that human activity, in this case burning of forests for agricultural purposes, can precipitate major health problems through air pollution. On an individual level, breathing the air there was said to be equivalent to smoking four packs of cigarettes per day.

### Other Acute Air Pollution Incidents

*Poza Rica, Mexico (1950)*

In this incident an oil refinery burning off excess hydrogen sulfide ($H_2S$) gas experienced a major $H_2S$ release when the flame was extinguished accidentally with no automatic cutoff in place.

*Bhopal, India (1984)*

This highly publicized incident occurred when a Union Carbide chemical factory accidentally released methylisocyanate into the surrounding community. With no effective zoning restrictions, thousands of people lived in immediate proximity to the factory, taking advantage of electric power hookups not otherwise available to run their television sets. More than 2,000 people died in the days immediately following the release and at least 5,000 more as a later result of the exposure. Thousands of others were made ill from this accident. Lung damage and damage to the immune system are blamed for an ongoing extremely high tuberculosis rate among the 500,000 residents affected. The followup included years of litigation to recover monetary damages.

*Cameroon, Africa (1986)*

An unknown gas (thought to be $CO_2$) was released by an eruption from the bottom of a volcanic lake. This is an example of a strictly natural phenomenon, and thus not easily preventable, which nonetheless resulted in numerous human fatalities in a location where emergency response capability was minimal.

### Airborne Infectious Diseases

Several infectious diseases can be transmitted via the airborne route. Tuberculosis is in that category although the great majority of TB transmission occurs person-to-person in crowded indoor conditions. Q fever is another example. A community outbreak of that disease associated with a sheep farm is described in Chapter 14.

The most significant airborne community infectious illness is legionellosis (Legionnaire's disease). First identified in 1976 following an American Legion convention in Philadelphia, legionellosis is a bacterial disease, with some 1,500 confirmed cases annually. It is also suspected as the agent of about one in seven cases of pneumonia. The disease occurs endemically and in outbreaks affecting primarily the elderly or immune-suppressed population. Its reservoir is in warm stagnant water. Community outbreaks have been traced, for example, to aerosolization from cooling towers and evaporative condensors atop hospitals or hotels. It is also found in hot-water tanks, shower heads, and room air conditioners indoors.

The organism has unique nutritional requirements including iron, cystine, and various nucleic acids. Although the organism is found in at least 40 percent of freshwater sources, its unique growth requirements are seldom available. Thus, it is now believed to be an intracellular parasite invading various protozoa, primarily amoebae, which can provide those unusual nutrients. Intervention strategies include the use of conventional biocides, or periodic superheating or superchlorination.

In the 1990s another example of human health effects from an airborne biological agent was described. The toxic dinoflagellate *Pfeisteria piscicida*, which has caused massive fish kills in the outer banks of North Carolina and in Chesapeake Bay, can cause neurological symptoms in humans exposed at close range to the aerosolized toxic agent.

## Chronic Health Effects of Air Pollution

Although numerous investigations have attempted to relate ambient air pollution to chronic respiratory illnesses such as bronchitis, asthma, emphysema, and lung cancer, this has proven to be an elusive goal. While comparisons of rural and urban populations show some tendencies toward chronic effects from higher air pollution levels, the studies are plagued by confounding factors. Tobacco use and occupational exposures are the most difficult to overcome. Even agricultural occupations include activities which potentially expose the population to high levels of pollutants such as grain dust, animal dander, and gasses from silage fermentation.

## Ecological Effects of Air Pollution

One of the most widely debated ecological effects of air pollution has been the phenomenon known as "acid rain." Actually, all rain is slightly acid, averaging 5.0 to 5.6 on the pH scale. The hypothesis is that $SO_2$, particularly from coal burning power plants, acidifies rain resulting in greater acidification of lakes. The problem has been particularly noticeable in the Adirondack region of New York State. Deleterious effects are seen as reduced fish populations, either directly from the low pH or from aluminum leached from soil by the acidic water. It is also postulated that the acidity can affect plant life dependent on affected soil. It is still unclear how much of the lake acidity is due to human activity rather than natural phenomena. The good news seems to be that acidity levels in lakes are stabilizing. The fact that $SO_2$ levels in ambient air have been decreased about 25 percent since the 1970 Clean Air Act imposed restrictions on emissions may be at least partially responsible for the stabilization.

Perhaps a more significant ecological effect has been that associated with smelting operations. The nickel smelting operation in Sudbury, Ontario, for example, was apparently responsible for decimation of most plant life in a several-mile radius around the smelter in the 1960s and 1970s. End-of-pipe controls have helped to improve that situation somewhat since then. Other smelting operations have had similar effects.

## INDOOR AIR QUALITY

For Environmental Health professionals, the 1980s and 1990s saw a significant rise in complaints and problems related to indoor air quality (IAQ). The increase corresponded to changes in building ventilation brought about by the energy crisis of the 1970s. New buildings were being built with sophisticated ventilation controls, central air conditioning, and unopenable windows. Individuals working in such buildings had little control over the environment. Work stress and the lack of control are contributing factors to *sick building syndrome* (SBS). Meanwhile, increasing energy costs prompted a reduction in fresh air intake and more recirculation.

The result has been increasing reports of SBS with numerous building occupants citing a combination of symptoms such as coughing, wheezing, headaches, dizziness, and others which apparently cleared up when they left the building. Specific solutions are often not apparent although sometimes the presence of workers monitoring for contaminants and making minor ventilation adjustments has brought about a reduction in complaints. This has been referred to as the "Hawthorne" effect. The term originates from an industrial setting in New Jersey where the effects of such activity on worker complaints were specifically monitored.

Later, the term *building related illness* (BRI) was coined to differentiate SBS from situations where a specific cause could be determined. The environmental factors identified include temperature (which seldom pleases everyone in a work setting), relative humidity (particularly very dry conditions), and a variety of contaminants including VOCs, bacteria, molds, endotoxins, animal dander, mites, and insect parts. Molds, in particular, often resulting from accumulated water or soaking of insulation material from water leaks, have been associated with specific respiratory problems. Sometimes an increase in carbon dioxide is used as an indicator of IAQ problems. A level of 1,000 ppm (0.1%) or about three times the ambient level suggests that building ventilation is inadequate.

A particular contaminant that has attracted a great deal of attention is secondary or environmental tobacco smoke (ETS). Now identified by the EPA as a cause of cancer and other respiratory illnesses, ETS is estimated to be responsible for more than 3,400 fatalities annually in the United States. Initially, separate facilities for smokers were provided with an increasing number of smokefree areas. Airlines first provided nonsmoking sections, then became totally nonsmoking. Many companies, buildings, and institutions have now become totally smokefree as societal attitudes have shifted to protecting people from involuntary inhalation of smoke from the burning of carcinogenic weeds.

## AIR POLLUTION CONTROL METHODS

### Legislative Approach

The previously described acute incidents affecting human health and the obvious smog engulfing cities in the 1960s prompted federal action to reverse a perceived, serious air pollution problem in the United States. The passage of the National Environmental Policy Act (NEPA) in 1970 created the Environmental Protection Agency (EPA) and paved the way for significant environmental legislation. The Clean Air Act of 1970 was one of the early and significant moves in that direction. Concentrating on the criteria pollutants, standards were set with provision for further reductions as technology and economic feasibility allowed. Enforcement and tightening of the standards were delayed during the energy crisis of 1973-79 when cleaner fuels, particularly natural gas, were in short supply. However, those problems, and the inevitable opposition of industry, were overcome to produce significant reductions in community air pollution in the ensuing decades.

Subsequent reauthorization and amendments have since reduced the allowable emissions even further. Major amendents in 1990 halved the limits for $NO_2$ and $SO_2$ and provided for regulation of some 100 new chemicals. It is estimated that in 1988 half of the U.S. population lived in counties exceeding the proposed limits in the revised standards. The goal for the year 2000 was to reduce that total to no more than 15 percent of the population. However, in the late 1990s there remained many cities where monitoring indicated that many of the newly regulated chemicals were present in ambient air at concentrations exceeding the proposed standards. Implications for human health remain unclear.

### Combustion Improvements

With much of the air pollution problem associated with combustion of fossil fuels, the selection of cleaner burning fuels becomes a major part of the intervention strategy. This is a clear example of prevention as opposed to end-of-pipe solutions. Natural gas is the cleanest burning of the fossil fuels, followed by oil, then coal (low sulfur anthracite coal is preferable to high sulfur bituminous coal) and peat. This hierarchy was threatened during the energy crisis of the 1970s when fears were raised that world supplies of recoverable oil and natural gas might be limited or possibly exhausted before the middle of the twenty-first century. Only coal appeared to present a stable future supply. Those fears have been allayed, at least temporarily, as we begin the new millennium, but they may well reappear in the future. The future of nuclear energy also enters the picture and will be discussed in Chapter 7.

In addition to selection of cleaner fuels, use of technology for more complete combustion—higher temperatures, refined air/fuel ratios and secondary burning chambers—has served to reduce emission of pollutants.

## Effluent Controls

End-of-pipe effluent controls have been necessary to bring stack emissions (sometimes referred to as stationary sources) into compliance with emission standards. These controls physically intercept the pollutants, remove them from the effluent airstream, and capture them for management as solid or liquid wastes. Examples of such end-of-pipe controls include:

### *Howard chambers*

Howard chambers are baffled collection chambers where the airstream velocity is reduced to allow settling of heavier particulates.

### *Cyclone separators*

Cyclone separators are devices that utilize centrifugal force to throw heavier particles out of the airstream for collection at the perimeter.

### *Baghouse filters*

Baghouse filters, as the name implies, utilize fabric spread over a large surface area for physical removal of particles. Efficiency increases as a cake of solid material builds up on the surface until the filter must be backwashed to collect the particles. They are particularly useful in situations creating very dusty effluents such as the grain industry.

### *Liquid scrubbers*

In liquid scrubbers the effluent air is passed through a liquid medium enabling the removal of pollutant gases which are literally washed out of the airstream.

### *Electrostatic precipitators*

Electrostatic precipitators are the most expensive controls to install, and are used for removal of very small particles. The airstream is given a positive charge before it is passed over negatively charged collecting plates, which then attract and remove the oppositely charged particles.

*Biological filters*

Biological filters are devices used for removal of specific volatile organic compounds (VOCs) and other odoriferous compounds. The gas is passed through a filter medium consisting of either naturally available materials (such as peat, compost, or wood chips) or inert materials (such as sand, glass, or ceramics). The natural materials provide a mixed microbial flora and nutrients. The organisms which can successfully digest the organic compound of interest proliferate and literally eat the contaminant. Sometimes specific bacteria, known to digest the contaminant of interest, are seeded onto the filtration medium.

**Controls for Transportation (nonstationary sources)**

Because internal combustion engines and transportation sources remain the most significant contributors to community air pollution, great efforts have been expended to deal with them. Some of those efforts can be summarized as follows:

*Catalytic converters (an end-of-pipe control)*

The major approach which emerged in the United States following the 1970 Clean Air Act was the use of catalytic converters on the exhaust systems of motor vehicles. They were mandated for model year 1975 and later vehicles. They use platinum, pallidium, or rubidium to reduce nitrogen oxides to nitrogen and also to oxidize hydrocarbons. An added benefit is that they required leadfree gasoline, thus also reducing emission of lead into the environment. This approach has worked quite well for more than twenty years. One drawback is the production of nitrous oxide (laughing gas), which is itself a greenhouse gas.

*Alternative fuels (a source control)*

A number of cleaner burning fuels have been proposed as substitutes for gasoline. They include:

a) Compressed natural gas, which some experts think is the logical fuel of the future, producing fewer pollutants and requiring relatively minor changes to engine design.

b) Propane, which is used in some vehicles, particularly small trucks, but has the disadvantage of being explosive.

c) Gasohol (a mixture of gasoline and ethanol), which is also used extensively and is actually mandated during the winter months in some cities. This approach is controversial; critics say that the benefits in air pollution reduction are marginal and that subsidization of ethanol production is a political payoff to corn-producing agricultural interests.

d) Electric cars are actually in limited production. Technical problems associated with producing such vehicles competitively have been numerous. High weight, poor acceleration and short range before the need for recharge have all slowed the introduction of electric cars. New advances in battery design, weight reduction, and improved range and acceleration are all proceeding, and there should be a brighter future for this option in the twenty-first century. California actually passed legislation requiring an increasing percentage of electric vehicles to be introduced by automakers starting in 1996. The state has had to back away from the initial requirement because it proved to be impractical. However, this approach remains a viable future option.

*Engine and vehicle design innovation*

Numerous changes have been proposed to reduce emissions. Smaller engines, lighter weight, aerodynamic design, stratified carburetion, and leaner burning air/fuel ratios have been introduced. Computer-regulated operation has also become standard. Following the energy crisis of the 1970s there was a major trend toward smaller, lighter-weight vehicles with a federal mandate to increase average gas mileage efficiency. The return of plentiful gasoline, actually less expensive when adjusted for inflation in 1998 than before the energy crisis, has reversed that trend. The preference for pickup trucks, conversion vans, four-wheel-drive and other sport utility vehicles has grown in the 1990s. Thus, engineering innovations to improve efficiency and reduce emissions are being balanced by the trend toward larger, less efficient vehicles. The history of reduced speed limits further exemplifies the current trend. The mandated 55 mph limit imposed in the 1970s to conserve gasoline has been replaced by 70-75 mph on rural freeways in the late 1990s.

*Vehicle inspection programs*

Many states have imposed mandatory inspection programs to assure that individual vehicles are not violating pollution control standards. While some view this as just another government intrusion into people's lives, statistics are available to show substantial reductions in criteria pollutants resulting from the program. Modifications are possible to answer critics. For instance, in Minnesota, vehicles under five years old are exempt, and the inspection program applies only in the seven-county metropolitan area of the Twin Cities under the theory that pollution is much less of a problem in rural areas. There is considerable sentiment in the legislature for eliminating the inspection program completely.

*Mass transit vs. private cars*

The United States is unique among industrialized nations in its dependence on, and encouragement of, private cars. In other countries, extremely high gasoline taxes discourage driving and favor public transportation. Such taxes have been vigorously opposed in the U.S. both by the automobile companies and by the public. The love affair of Americans with their vehicles is legendary. The arguments of convenience and independence engendered by individual autos are difficult to counter. Meanwhile, the freeway construction and suburbanization that have encouraged the trend have brought about increasing traffic congestion, thus actually becoming a major factor forcing reappraisal of mass transit as a viable option. In cities such as New York, where geography and congestion have restricted suburban sprawl, mass transit is more successful. Many metropolitan areas continue to push for, and experiment with, light rail, park-and-ride transit hubs, express lanes for buses and carpools, and other ways to counter congestion and reduce air pollution. The twenty-first century will undoubtedly see continued innovation along those lines.

## REFERENCE

An historic report detailing the investigation of the Donora, Pennsylvania, acute air pollution incident of 1948 is:

U.S. Public Health Service. 1949. "Air Pollution in Donora, PA: Epidemiology of the Unusual Smog Episode of October, 1948." Public Health Bulletin No. 306.

# Chapter 6

# SOLID AND HAZARDOUS WASTE

In addition to the liquid and gaseous pollutants discussed in the preceding chapters, communities continue to wrestle with the problem of solid waste management. Unlike wastewater, which can be collected through a system of mains for central treatment, or gaseous discharges, which are simply put back into the atmosphere, most solid wastes must be collected and transported manually, increasing the costs and the potential for exposure of workers. The exceptions are individual incinerators (converting solid waste to air pollution), which have largely been rendered obsolete, and garbage grinders, which convert readily biodegradable solid waste into liquid waste but are restricted to a small percentage of the total waste generated. Thus, it is important to define the categories of solid waste to select the most efficient management system available for each.

## SOLID WASTE CATEGORIES

### Community Solid Waste

Community solid waste includes all nonliquid waste except bodily discharges. The general term *refuse* is used for all solid waste. *Garbage* is specifically the readily degradable, or putrescible, portion of the solid waste. It includes food wastes and grass clippings. Garbage has the advantage of decaying rapidly. However, it also presents the most immediate threat to human health, because it often harbors pathogens and attracts vermin, which are vectors of human disease. It also is the most unsightly and is associated with the worst of odors. *Rubbish* is the term used to describe the nonputrescible portion of the waste, such as metals, glass, and plastics. Paper and cardboard make up a high percentage of community solid waste and are intermediate between the truly putrescible and nonputrescible materials although they are usually treated as rubbish. Additionally, communities must manage some large-volume waste materials such as construction and demolition debris and trees and branches felled by windstorms.

### Industrial Solid Waste

The wide variety of industrial processes can generate atypical wastes that must be managed separately. Some of these wastes are *hazardous waste,* which I will define in the next section. However, industrial solid waste also may include large volumes of off-specification products or unconventional solids, such as sludge from waste-water treatment, or solids remaining from end-of-pipe air pollution devices. Industrial solid waste is often consigned to separate approved landfills even when not categorized as hazardous.

### Hazardous Waste

Prior to 1976, hazardous chemicals were routinely managed similarly to community solid waste. The waste frequently ended up in landfills subject to future development. In 1978, a highly publicized incident in upstate New York called attention to the problems of that practice. For many years, the Hooker Chemical Company, now a subsidiary of Occidental Petroleum, had deposited chemical waste—including large quantities of dioxin-containing chlorinated hydrocarbon pesticides—in a landfill site called Love Canal. The site, actually fairly well contained, was eventually filled to capacity, and in 1953 it was covered with compacted soil. It was then sold to the Niagara Falls School Board at their request, accompanied by a warning from the company regarding the hazardous chemical deposits. Eventually, a school and several hundred homes were built adjacent to the canal. Leaking chemicals and suspected health problems among the residents resulted in evacuation and relocation of the residents, as well as eventual substantial monetary compensation by Occidental Petroleum despite their warnings at the time of sale and controversy over the extent of human health effects.

The need for federal legislation to govern hazardous waste was addressed through the Resource Conservation and Recovery Act (RCRA) passed in 1976. The enforcement responsibility was given to the EPA. Hazardous waste was defined in the Act to include all chemical waste in the following categories:

a) *Corrosive* (pH $<2$ or $>10$).

b) *Ignitable* or *flammable* (flash point >60°C). Such materials can ignite through friction, absorption of moisture, or spontaneous chemical change.

c) *Reactive* (either shock or water sensitive).

d) *Toxic, Carcinogenic,* or *Teratogenic.*

RCRA has been termed "cradle-to-grave" legislation because it requires tracking and manifesting of these chemicals up to the point of final treatment and deposition or dispersion. The idea is to establish financial liability for any cost involved in such management and for future problems.

## Medical Waste

Although it constitutes less than 1 percent of the total solid waste stream, medical waste has attracted a disproportionate amount of attention. It was originally included in the RCRA, but the EPA did not establish rules for its management because the agency did not discern any specific hazard to the public unique to medical waste. In 1988, a series of beach washups of what appeared to be medical wastes resulted in considerable media attention. (It was later ascertained that most of the material actually came from individual residences in several Atlantic Coast states.) The concern arose from public fear of AIDS and the scenario that medical sharps (needles and syringes) might be contaminated and infect bathers.

The incidents prompted a temporary federal bill called the Luken Bill, which applied only to the affected states and expired in three years. Other states chose not to be included mainly because of the onerous tracking and manifesting provisions. However, many states passed independent legislation of some kind. Although state laws differ widely, a common theme is to segregate infectious waste (microbial cultures, sharps, blood and other body fluids, infected research animal carcasses and pathological specimens) for separate handling, often referred to as "red bag" waste. The major difference in handling is that red bag waste must be decontaminated (by heat or chemical treatment) and rendered unrecognizable before joining the community solid waste stream.

Medical institutions have had to overcome extreme community opposition (referred to as the "not-in-my-backyard," or NIMBY, syndrome) to site such management facilities. Untreated medical wastes have been specifically excluded from most landfill, mass burn, and refuse derived fuel (RDF) facilities. The irony is that there has been no evidence of infection from medical waste spread to the community. However, in 1997 there was a report from the state of Washington of the possible occupational transmission of tuberculosis (3 cases and 13 conversions) to employees of a commercial medical waste facility. Apparently, filters on the shredding device became clogged, resulting in the aerosol bypassing the filter and infecting the workers. Ideally, shredding should be carried out only after decontamination by steam.

## Incinerator Ash

The final category of solid waste is incinerator ash. There has been some controversy as to whether such ash should be handled as hazardous waste at considerable extra cost, or whether it can safely join the community solid waste stream for land deposition. In reality, the ash should not be hazardous under the RCRA definition as long as communities take reasonable precautions to keep significant amounts of hazardous chemicals out of the waste stream being incinerated. It has also been proposed that incinerator ash could be recycled as roadfill material, a suggestion which has met with considerable opposition.

## SCOPE OF SOLID WASTE GENERATION

If sporadic large-volume materials such as demolition debris and storm damaged trees are excluded, the current estimate is that in the U.S. we generate about 4 pounds of solid waste per person every day. A major goal, regardless of the management method utilized, has been to reduce that volume to no more than 3.6 pounds per person per day by the year 2000. Waste minimization is driven by the high cost of storage, collection, and "disposal." In reality, the term *disposal* is inappropriate, because if waste is not recycled or reused, the end product of treatment and transport is either dispersed into air, water, land, or temporary (pending decay) or permanent storage. Cost estimates in 1998 dollars are approximately $50 per ton for landfilling or for incineration. The costs for RCRA-regulated wastes are about four times higher.

## PUBLIC HEALTH SIGNIFICANCE OF SOLID WASTE

Selection of strategies and siting of solid waste management facilities has been driven largely by the considerable economic and aesthetic concerns associated with the problem rather than public health implications. While some occupational health problems are associated with handling solid waste, there is generally no direct pathway of public exposure. Rather, the health concerns stem from indirect exposure to air or water that has been contaminated through faulty solid waste management practice. In addition, there is the potential for vector-borne diseases, stemming from inadequate practices allowing vermin to infest putrescible solid waste. Those diseases will be discussed in Chapter 14.

## MANAGEMENT STRATEGIES FOR SOLID WASTE

Steps in the management of the community solid waste stream, to satisfy aesthetic, economic, and public health concerns, must include a combination of volume reduction, recycling or reuse, and transport, followed sometimes by treatment processes and ultimately by long-term storage (in landfills) or dispersal of the remaining waste into the air or water.

### Waste Reduction or Minimization

In the United States we have developed what many consider to be a very wasteful philosophy of resource exploitation, based more on convenience than on conservation. For example, we use an enormous amount of packaging material (which inevitably joins the solid waste stream) even for very small items. This is sometimes justified on the basis of making shoplifting more difficult. We also have increasingly shifted from reusable to single-use items for everything from infant diapers and medical supplies to cameras. The shift is often justified on economic grounds as well as convenience—and for medical supplies, greater assurance of sterility and, therefore, greater patient safety. On the other hand, it is often not clear whether the economic calculations include the cost of solid waste management. The controversy is particularly difficult for environmental health professionals because the benefits to public health and safety of single-use products often conflict with the environmental degradation effects of voluminous solid waste.

**Recycling and Reuse**

*Community recycling (paper, glass, metal, plastic)*

From an environmental management perspective, the worthwhile goal of recycling materials or reusing items has been emphasized and increasingly met. Nationally, more than 20 percent of the solid waste stream, and in some locations, more than 40 percent, is now recovered. Unfortunately, recycling practices are generally more labor intensive, and present greater occupational health hazards, than other waste management practices.

Community recycling programs have progressed considerably since the early efforts of the 1970s. The key has been to develop markets for recycled material and to make the collection systems more efficient. The early reliance on citizens bringing recyclable items to a collection center has been replaced by organized community pickups, often with monetary incentives in the form of reduced solid waste charges on the utility bill. Householders have responded with a willingness to segregate items accordingly. Recycling includes newspapers, cardboard, metal cans and containers, glass items, and, increasingly, many types of plastic. The recycling symbol is stamped onto those plastic containers that can be recycled.

Currently, the practical maximum waste stream recovery for communities is thought to be about 50 percent of the waste stream. There are detractors who claim that some recycling efforts are misdirected and are neither cost effective nor environmentally beneficial. One argument compares single-use styrofoam cups to reusable ceramic coffee mugs and concludes that the ceramic mug would have to be used more than 1,000 times to be more cost-effective than styrofoam cups when considering the use of water and detergents to wash the mug and the environmental impact of that water and detergent use.

*Composting*

The conversion of putrescible wastes into fertilizers for soil enrichment via composting has long been practiced in some communities and by individuals. The practice has been well accepted in areas where conventional fertilizers are expensive. In the United States, most communities have concluded that the cost of composting is not competitive with other fertilizers. However, many communities now collect yard waste specifically for composting. The process involves careful temperature control and frequent mechanical turning of the materials for aeration. Aerobic bacteria act on the putrescible material to convert it into fertilizer.

*Feeding of garbage to hogs*

One of the earliest recycling methods practiced in many communities was the collection of kitchen wastes from restaurants or institutions by local hog farmers. This was a mutually beneficial practice for both parties. Farmers were getting feed for their hogs for the cost of their time and labor to go into town for it. Institutions were getting rid of part of their waste stream without cost. From an aesthetic and public health perspective, however, the weekly or biweekly frequency of pickup was not usually sufficient to prevent odors and vermin infestations, particularly during hot weather. More importantly, the practice was seen as one means of transmitting trichinosis to the hogs. Trichinosis is a helminthic disease acquired by hogs and humans from ingesting inadequately cooked meat containing encysted worms (see Chapter 12). The hog feeding practice was greatly curtailed in the 1950s after laws were enacted requiring that garbage intended to be fed to hogs be heat pasteurized. The increased cost to farmers made the pasteurization practice noncompetitive with other feed arrangements.

*Heat energy recovery*

Sometimes overlooked is that solid waste can be "recycled" through the process of heat energy recovery from the incineration of the material. This recovery has become a major feature of waste-to-energy mass burn incinerators and of refuse derived fuel (RDF) facilities, which will be discussed later in this chapter.

**Water Carriage Systems**

The installation of garbage grinders in homes, restaurants, and institutions is common practice as a waste management system. The grinders convert solid waste to a slurry which then joins the wastewater stream in sanitary sewers for transport to the treatment plant. The advantages of grinders are many. They provide immediate transport to a treatment facility without the sanitary and aesthetic problems of storage and without the cost of over-the-road hauling. A major disadvantage is that grinding applies only to putrescible kitchen wastes and therefore only a small portion of the solid waste. In addition, the waste has a high BOD and imparts an additional load on the wastewater treatment facility. In areas served by septic tanks rather than municipal sewers, grinders can be a cause of system failure. One other problem associated with grinders is the potential proliferation of rodent infestations in the sewers. The garbage provides a convenient food source for rats, leading to population explosions that can spill over to surface property. The implications for pest management will be discussed in Chapter 15.

**Land Disposal of Solid Waste**

Approximately 75 percent of the nonrecycled community solid waste in the United States ends up being stored in landfills. Historically, land disposal for many communities took the form of open dumps. It was a matter of "out of sight, out of mind." When suitable sites were available in sparsely populated areas close to town, the town dump was an inexpensive means of managing solid waste. The obvious problem was that such dumps were environmental disasters. Groundwater contamination, particulates from spontaneous combustion, vermin proliferation, odors, and aesthetics, and scavenging by people and pets were all inevitable.

By the middle of the twentieth century, open dumps were being replaced by sanitary landfills. The concept is relatively simple. The solid waste, requiring little or no segregation of materials, can be spread over low-lying land or into prepared trenches by bulldozers. The waste is compacted daily and immediately covered with excavated soil. When managed properly, most of the previously mentioned problems are alleviated. Seepage into groundwater remained a problem until liners and leachate collection systems were added. Methane gas formation and seepage to the surface has been reported, although that problem can be turned into another recycling asset with methane recovery. Marginal (low-lying or marshy) land, with no other obvious use can be reclaimed for useful purposes once filled with solid waste and compacted. Numerous airports (including those in New York City), parks, and golf courses were created out of landfill sites.

With increasing suburban development, suitable sites have become scarcer and more remote. Long hauling distances add to costs, and the NIMBY reaction has made site selection very difficult if not impossible in some metropolitan areas. These difficulties have contributed to the push for waste reduction and recycling to decrease the volume of waste going to landfills and extend the life of those in use.

An additional problem encountered by landfilling is that of long-term liability. Although some putrescible material rapidly decays and returns to the soil, nonputrescibles can remain in permanent storage for many decades. Legal responsibility can remain perpetually for the waste generators.

Before the enactment of hazardous waste legislation, toxic chemicals were often deposited in landfills, and liability issues persist to this day. The Superfund legislation added to the RCRA in 1980 was intended to address the problem and will be discussed in a later section of this chapter.

**Mass Burn Incineration**

About 15 percent of the nonrecycled solid waste is now managed by waste-to-energy mass burn incineration. The growth of this method has been greatly slowed by extensive opposition from community groups. There are numerous advantages to municipal incineration. Relatively little land is required, so facilities can be built centrally, often in industrial areas near downtown. The central location reduces transport costs. When an incinerator is operated efficiently, waste is quickly deposited into the burning chamber, minimizing sanitary problems. When the heat recovery potential is extensively utilized, power may be provided for downtown commercial buildings. With effective recycling programs and a means to keep hazardous materials out of the waste stream, additional segregation is unnecessary. Most importantly, waste volume is reduced by 85 to 90 percent.

A main objection has been the potential air pollution problems, which can result from dispersion of incinerated solid waste (particularly heavy metals such as mercury and lead) into the air. Stringently worded permits spell out the emission limitations for these facilities. Expensive end-of-pipe controls, particularly electrostatic precipitators, are needed to comply with the rigid standards. Additionally, the problem of what to do with the remaining ash has created controversy. As previously mentioned, some advocate treating it as hazardous waste at a greatly increased cost for landfilling, and others propose using it as roadfill material.

### Refuse Derived Fuel

One quite successful management strategy for solid waste has been the more recent creation of refuse derived fuel (RDF) facilities. In such facilities, about 65 percent of the waste is suitable to be pelletized for burning in electric power operations. Another 25 percent can be recovered as recyclable material (including heavy metals), and the remaining 10 percent is landfilled. The disadvantage is that the operation is more labor intensive and occupationally dangerous due to the sorting operations.

## MANAGEMENT OF HAZARDOUS WASTE

The 1976 RCRA to regulate the management of hazardous waste has been described. In 1980 the Comprehensive Environmental Responsibility Compensation Liability Act (CERCLA) added the Superfund provision to hazardous waste management. Superfund was established to pay for the cleanup of years of unregulated disposal practices of the RCRA chemicals. It was designed to collect the cleanup costs from responsible parties, although often those responsible were out of business, or had merged or been bought out by other entities, making establishment of responsibility a drawn-out, contentious, and complex legal process. The "deep pockets" concept applies to wealthier companies and government agencies which are deemed able to pay. Government funds (the "Superfund") are made available when no other viable source can be identified. The Superfund Amendments and Reauthorization Act of 1986 (SARA) added underground storage tanks to the regulations, and later reauthorizatons have been aimed at reducing litigation while involving local community interests in the process to a greater extent.

The strategies available for actual management of the hazardous wastes are similar to those for community solid waste. They include volume reduction, recycling, neutralization, incineration, landfilling, and reforestation.

*Volume reduction* is accomplished by substitution of less hazardous chemicals in industrial processes and reuse of unused chemicals for other purposes as an alternative to disposal. *Recycling* includes recovery of valuable components (such as mercury) from industrial processes. *Neutralization* can be accomplished, for instance, by raising or lowering the pH of strong acids or bases to render hazardous chemicals non-hazardous. Bioremediation is also in this category. Genetically engineered microbes which can specifically degrade target hazardous chemicals to innocuous compounds are increasingly used to remediate hazardous waste sites.

Specifically designed *incinerators* utilizing high temperatures and secondary burning chambers are available selectively for hazardous wastes. Some toxic solvents can actually be used as fuels when burned in these facilities. Often, portable heavy duty incinerators are transported to Superfund sites for incineration of contaminated soil. Heavy metals are generally excluded from even the most sophisticated incineration processes.

*Landfills* for hazardous waste are greatly restricted and must be designed with strict containment provisions such as liners and leachate collection systems. In some instances, Superfund sites are actually remediated by physical relocaton of contaminated soil from one location to a specially designed landfill located elsewhere.

An innovative approach to remediation of contaminated sites is *reforestation,* in which natural processes are set in motion by planting trees in the contaminated soil. It works via several mechanisms. For example, enzymes encourage bacterial growth, a means of bioremediation called *phytostimulation.* Also, contaminants can concentrate both in the leafy growth of the trees, where they are broken down by photosynthesis (*phytotransformation*), and in woody root systems (*phyto-accumulation*). None of these processes causes any apparent permanent damage to the trees themselves.

# Chapter 7

# IONIZING AND NONIONIZING RADIATION

Of all the environmental health issues that are being publicized at the turn of the millennium, the one most perplexing to the general public is ionizing radiation. While dirty water, smog laden air, and odoriferous garbage frequently elicit the "outrage factor," at least they can be seen and/or smelled. Thus, the public can react to something they know is there and is aesthetically displeasing. Radiation has a mystique all of its own. Invisible and insidious, it conjures up memories of the cold war days when the threat of nuclear annihilation hung over the world. Nuclear accidents such as the Three Mile Island near meltdown in Pennsylvania in 1979 and the Chernobyl disaster in the Ukraine in 1986 have served to reinforce the reality of those fears.

Thus, at the turn of the century, those responsible for environmental risk communication are being challenged to explain current realities and dangers of radiation to a still skeptical public. A classic example of this task is the reluctance of the public to accept irradiated food even in the face of increasing threats of food-borne illness. That issue will be discussed further in Chapter 12.

What risk does radiation pose to the public in the United States and elsewhere? In a way, explaining radiation is like trying to explain electricity. We all know electricity exists (and indeed we feel quite helpless when temporary power outages deprive us of its use), but we are dependent on power company engineers to harness, generate, and deliver it into our homes. Technically, ionizing radiation (ionization) refers to elements containing sufficient energy to release electrons from the atom. These elements "decay" as the electrons are released and ultimately become stable new elements. The time required for this process is variable, ranging from seconds to thousands of years, but ultimately all earthly elements will no longer be radioactive. Meanwhile, human health effects result when ionization occurs in living tissue.

## TYPES OF RADIATION

There are three basic types of naturally occurring radiation:

### Alpha

Alpha radiation is the only type which actually has mass. It consists of particles with two protons and two neutrons. Relatively nonpenetrating, it can be stopped by a sheet of paper or the skin on a human hand. It is dangerous because radioactive particles can be absorbed in soft tissue (such as the human lung) and continue to ionize. Radon, which will be discussed later in this chapter, is an example of an alpha radiation problem of public health importance.

### Beta

Beta radiation consists of electrons emitted from the nucleus of an atom (not orbiting electrons). Examples include isotopes of nitrogen and sulphur, carbon-14 (C-14), and tritium (H-3), many of which are used as tracers in scientific research. Beta has intermediate penetrating power but would be stopped, for example, by a half-inch-thick sheet of lucite. Because it has no mass, there is no continuing effect on an individual beyond the original absorbed dose.

### Gamma

Gamma radiation consists of photons emitted from the nucleus. Examples include cobalt-60, iodine-125, and strontium-90. Gamma is the most penetrating form of radiation and requires thick lead or concrete shielding. Because it has no mass, items which have been irradiated with gamma rays such as cobalt-60 (such as irradiated food), are not radioactive, and do not pose a radiation threat to humans. The absorbed dose is sufficient to destroy pathogens on or in the food, but the irradiated food poses no health threat upon ingestion.

### X Rays

X rays are a man-made form of radiation, similar to natural gamma rays. They consist of photons emitted from the nucleus as the result of deceleration of high-energy electrons in high-atomic-number materials (such as tungsten). When X-ray machines are turned on they produce an absorbed dose to the tissue being X-rayed. However, once turned off, they pose no further health risk.

## UNITS OF MEASUREMENT

The variety of units of measurement associated with ionizing radiation is confusing to the public and contributes little to understanding. Changes in the terminology from familiar to less-familiar terms has added to the confusion. The following are some of the terms currently in use which I believe are important to understanding the human health implications of radiation:

### Half-life ($T_{1/2}$)

This is the time required for 50 percent decay of the element that is releasing electrons from its atomic structure. Half-life varies from seconds to thousands of years, and eventually will result in extinction of radioactive materials. Meanwhile, materials with a long half-life are at the core of the debate on management of nuclear wastes, which will be discussed later in this chapter.

### Radiation Absorbed Dose (rads)

This is the term which describes the radiation energy absorbed by a human when exposed to ionizing radiation. A rad is defined as 100 ergs of energy/gram of tissue.

### Roentgen Equivalent Man (rems)

This term describes the biologically effective dose, recognizing the importance of the form of energy absorbed as well as the amount. It is expressed in terms of rads multiplied by a quality factor (rems = rads x quality factor). The quality factor adjusts for the relative hazard. For example, for alpha particles the quality factor is 20, which takes into account the continuing ionization in human tissue due to the presence of the particles. By contrast, gamma and X rays do damage only in passing through the tissue and are assigned a quality factor of 1. Table 7-1 lists other terms which are used to describe levels of radiation.

Table 7-1

Radiation Conversion Factors

| Term | Conversion Equivalent or Definition |
|---|---|
| Bequerel | 1 disintegration /second |
| Curie | 3.7 x $10^{10}$ bequerels |
| Gray | 100 rads |
| Seivert | 100 rems |

## HEALTH EFFECTS OF RADIATION

The effects of radiation can be somatic (biological effect on the organism) or genetic (effect on future generations). Table 7-2 lists the somatic effects expected from various doses of absorbed radiation.

Table 7-2

Dose Effects on Human Health

| Dose Absorbed (rems) | Expected Effect |
|---|---|
| 25-100 | transient reduction of lymphocytes or neutrophils |
| 100-200 | nausea and fatigue, usually full recovery |
| 300-600 | decrease in white cells, 50% fatalities in 30 days |
| >600 rems | death, usually within 2 weeks |

By contrast the average exposure of U.S. citizens is measured in millirems (1/1000 of a rem). Table 7-3 lists the average exposure by source.

Table 7-3

Average Exposure of U. S. Citizens

| Source | mrem/person/year | Notes |
|---|---|---|
| Natural background | 300 (82%) | radon, cosmic rays, body materials |
| Medical | 53 (15%) | diagnostic (X- rays) and therapeutic |
| Occupational | 0.9 (<1%) | |
| Power plants | 0.05 (<1%) | |
| Miscellaneous | 10 (2.7%) | jet airplane flights, color TV sets |
| Total | 364 | (about 1 mrem/day) |

The public limit recommended by the Nuclear Regulatory Commission (NRC) is 500 mrem per year. Clearly, the individual exposure will vary depending on many factors, particularly occupation, geographic location, and residential radon level. The occupational limit is set at 5,000 mrem, or ten times the public recommendation. The reason for that discrepancy is that occupational exposure sources are known and monitored on a continuing basis, often with personal dosimeters, while the public exposure is not. Thus, workers who receive unusually high doses or are approaching the occupational limit can be treated or relieved of duties to assure that they are not overexposed.

## OCCUPATIONAL EXPOSURES

There have been some occupational exposures recorded in an anecdotal manner which are informative. They include the following:

### Radium Dial Workers

Early in the twentieth century, many wristwatches were manufactured with fluorescent dials so they could be seen in the dark. Radium-based paint was used for that purpose before people knew of its highly radioactive properties. Women workers who applied the paint did so manually, using thin-tipped brushes which they kept sharp by licking the tips. The inevitable result was extremely high rates of bone cancer and leukemia, providing definitive evidence of the radiation hazard.

### Uranium Miners

A second example of very high occupational exposure is provided by the experience of uranium miners. Fully 20 percent of the miners in this industry, working prior to meaningful occupational health regulations, have died or will die of lung cancer. This situation is confounded by a very high smoking incidence among this group of workers. Although significant monitoring was not carried out until later, it is recognized that the earlier miners were exposed to extremely high levels of radon. The current EPA recommendations on residential radon are actually extrapolated from the uranium mine experience.

### Other Occupational Exposure

Certain other occupations exposed to ionizing radiation such as radiologists, radiation technicians, dentists, and cyclotron workers have been reported to have higher than normal rates of leukemia, skin cancer, and cataracts. It is expected that current protective measures should reduce that excess risk to near zero for workers in those occupations.

## ACUTE PUBLIC EXPOSURE TO RADIATION

Since the dawn of the nuclear age, our knowledge of high-level exposure of the public to radiation is based primarily on the atomic bombing of Hiroshima and Nagasaki in August 1945 and the serious nuclear power plant accident in Chernobyl in the Ukraine in April 1986. The atomic bombs dropped on Japan killed several hundred thousand people, mostly from the extreme heat and subsequent fires from the blasts. The number of subsequent deaths from radiation effects is not certain, but fatalities from leukemia and other radiation-induced cancers continued for many years after the bombing. Genetic effects on future generations have also been studied and may be less serious than initially expected.

The Chernobyl incident in 1986 was the most serious nuclear power plant accident of the nuclear age. Again, no one knows the exact number of fatalities, but it has been reported to be as high as 150,000. Ongoing effects, particularly thyroid cancer in children, has been reported to be very high. Nuclear debris (including long-half-life isotopes such as uranium-238, strontium-90 and, cesium-137) also contaminated soil for miles around the explosion, leaving that land unsuitable for agriculture for many decades.

### Residential Radon

For the general U.S. public, residential radon, seeping into basements from natural sources in surrounding soil and rock, remains the most significant source of potentially hazardous radiation. radon, mostly in the form of radon-222, is a decay product of uranium-238 which decays to thorium, then to radium, then to radon-222, a noble gas which attaches to dust particles in the air. There are also short-lived "daughter" products such as lead, bismuth, and polonium, which are also forms of alpha radiation. Radon gas enters homes through cracks or crevices in basement walls or through water seepage. Radon-222 is a form of alpha radiation. The particles attach to dust which is inhaled by residents.

Actual exposure is a function of time spent in the basement and activity level. A person who spends considerable time in a basement recreation room, exercising vigorously, will be exposed to a much higher dose of radon than someone who rarely visits that level of the house. Upper floors are generally at much lower radon concentrations than the basement. The EPA has set an action level (level for remedial action) at 4 picocuries/Liter of air (150 Bequerels/$m^3$). This limit is extrapolated from data available on cancer rates of uranium miners exposed to much higher levels. The recommendation remains quite controversial. EPA has estimated that up to 15,000 cases of lung cancer annually might be caused by residential radon. Others dispute that figure because it assumes extensive occupancy of the area in which the 4 picocuries/Liter concentration is exceeded.

Nonetheless, monitoring devices are available to determine the level in individual basements. The level can be highly variable from home to home even in a single neighborhood. In some areas more than two out of three homes have been found to exceed the 4 picocurie limit at which remedial action can be taken. Additional ventilation can be achieved by the installation of exhaust fans, and cracks and crevices can be sealed to keep out contaminating air and water.

## CONTROL STRATEGIES FOR LIMITING RADIATION EXPOSURE

The basic control strategy for limiting exposure to radioactive sources is the triumvirate of time, distance, and shielding. Given a constant dose rate, the dose received is in direct proportion to the time of exposure. The dose rate is also inversely proportional to the square of the distance from the source (inverse square law). Finally, physical barriers can shield an individual from exposure. As mentioned earlier, the barrier needed is dependent on the type of radiation. A single sheet of paper is sufficient for alpha radiation, 0.3 inches of lucite is needed for beta rays, and 2.5 inches of lead for gamma or X rays.

## COMMERCIAL APPLICATIONS OF IONIZING RADIATION

### Nuclear Power Industry

In some countries, particularly in Europe, nuclear energy provides most of the electric power. In the U.S. it provides about 10 percent. Advocates cite the low pollution levels, competitive costs, abundant fuel supply, and relative safety of the operation. Opponents cite the unsolved issue of nuclear waste management (see next section of this chapter), and accident potential exemplified by Chernobyl and Three Mile Island. The fact is that in the Three Mile Island incident, the containment system prevented a meltdown, and public exposure to radiation from that incident was minimal.

While the Chernobyl accident was far more serious, the Soviet nuclear plant design is significantly different from plants built in Europe or the United States, and the nuclear industry claims that such an accident is exremely unlikely to occur in properly designed facilities. As of 1998, there were no nuclear plants either under construction or planned in the United States. The fear quotient remains very high, and antinuclear activists are a very potent force in preventing the construction of such facilities.

### Food Irradiation

The irradiation of food for preservation and killing of pathogens has been practiced for many years and is approved for a wide variety of items including spices, grains, poultry, pork and most recently, ground beef. The process utilizes cobalt-60 (or sometimes cesium-132) and very effectively destroys pathogens. However, opponents voice opposition and the public has been slow to accept irradiated food. Some fear that the food is radioactive, although that is not true. It has also been claimed that the radiation dose affects chemical bonding, resulting in loss of nutritive value and vitamin destruction, as well as the creation of radiolytic by-products. Most scientists believe that such reactions are minimal and dismiss the objections as trivial relative to the benefits to food safety.

### Sterilization of Single-Use Medical Products

The medical products industry has been much more successful than food irradiators in selling the idea of gamma sterilization of single-use products. Again, cobalt-60 irradiation is the process most widely used, and again, the product does not become radioactive. This process has contributed to patient safety and has proven economical for the healthcare industry. Public and activist opposition has been much less than for the food industry, probably because the products are not directly consumed and the mechanism of sterilization is not widely publicized.

### Scientific Research and Nuclear Medicine

Radioisotopes (primarily carbon-14 and tritium) are also used as tracers in scientific research. Iodine-125 and other isotopes have long been used in therapeutic medical applications, particularly as a cancer treatment. These medical and scientific applications have been widespread, with relatively little opposition, but they contribute to the radioactive waste problem to be discussed in the next section.

## RADIOACTIVE WASTE MANAGEMENT

Radioactive waste management remains a largely unsolved and thorny problem which has undoubtedly slowed the growth of the nuclear industry. It is regulated at the federal level by the Nuclear Regulatory Commission (NRC). Radioactive waste is divided into two categories, high-level waste and low-level waste.

High-level radioactive waste includes spent fuel from nuclear power plant cycles, (uranium-238, strontium-90 and cesium-137), tailings from mining and milling nuclear fuels, debris from dismantling obsolete nuclear power plants, by-products of breeder reactors and waste from nuclear weapons production (which is controlled by the military).

Low-level radioactive waste is all other radioactive waste not defined as high level. It consists primarily of radioactive waste from the scientific research and nuclear medicine segments of the industry. Low-level radioactive waste is generally lower energy and less hazardous, and has half-lives measured in hundreds of years or less, compared to thousands of years for some high-level waste.

### High-Level Radioactive Waste Management

Because high-level nuclear waste consists primarily of long half-life materials which may remain hazardous for literally thousands of years (10,000 years is often cited as the benchmark), the only viable management strategy is one of permanent concentration and containment. Much of the controversy over this issue stems from the difficulty of assuring such indefinite containment.

Not surprisingly, politicians are reluctant to be associated with any involvement of their jurisdiction even if there are financial incentives. The issue hinges on the ability of the federal government to find a permanent repository for the waste. Legislation passed in 1982 promised that the federal government would take responsibility for this waste and promised a site by 1998. That clearly proved unfeasible and the date has been set back well into the twenty-first century, leaving the industry scrambling to find temporary storage until a permanent site is available.

At temporary sites most of the waste is stored in above-ground casks. It is often vitrified (solidified into a glasslike structure) to minimize the leak potential and allay fears of accidental release during transport, which will eventually be needed if and when a federal site becomes available. In early 1999, the U.S. Congress agreed to assume federal liability for the temporarily stored high-level waste until a permanent storage site is available.

Much of the effort in finding a permanent storage location has centered on a site at Yucca Mountain, Nevada, but other sites are also being considered. Environmentalists question the stability of underground storage near earthquake prone fault lines. Environmental justice issues are also being raised whenever American Indian reservations are considered.

**Low-level Radioactive Waste Management**

Some low-level radioactive waste also consists of long half-life material requiring indefinite containment. Other strategies for short half-life wastes include the "dilute and disperse" principle whereby the waste can be discharged into sanitary sewers or incinerated if it is sufficiently diluted to meet emission standards. It is not surprising that environmentalists frequently oppose that option. Another strategy is that of "delay and decay," whereby shorter half-life waste can be held in storage until such time as it is no longer considered radioactive, then managed as nonhazardous waste.

Even low-level long-term containment facilities are in short supply in the U.S. The federal government has affirmed that such sites are not obligated to receive waste from out-of-state customers indefinitely. States are basically responsible for the waste they generate, although many states have entered into regional compacts to seek joint solutions to the problem. A few regional sites are available and financial incentives are persuading some suitable facilities to accept waste from other states. Long-distance transport remains an issue, with concern over accidental spillage along convoluted routes.

## POTENTIAL FUTURE ENERGY SOURCES

What energy sources may be available to industrial societies in the next millennium? As we have seen, fossil fuels, although currently appearing adequate, are finite, and shortages are inevitable at some time in the future, even for the most pollution-prone fuels such as soft coal and peat. Conventional nuclear power has advantages, but unless the long-term waste storage issue can be resolved, will remain stagnant, at least in the United States. Breeder reactors are considered too dangerous because of their use of plutonium.

Nuclear fusion (using sea water as a fuel and producing no radioactive waste) remains the best long-range hope in the minds of many, but progress in developing a feasible fusion process has been frustratingly slow. It can only be hoped that such a process can be developed before the fossil fuels are exhausted. Renewable energy sources such as solar, wind, and hydroelectric power have been limited by cost and dependability but remain as options which will probably increase as a percentage of the market as other options fail. Hydroelectric power is currently a very viable solution in areas where suitable conditions exist. It will be interesting to see how this scenario plays out in the twenty-first century. Increasing energy demand, depletion of fossil fuel reserves, ecological considerations, and unresolved nuclear waste issues will test concepts of "free market" solutions versus "net societal benefit" to the fullest.

## REFERENCES

Several references which provide a good overview of the problem of residential radon are as follows:

Archer, V. E. 1991. A review of radon in homes: health effects, measurement, control, and public policy. *Appl. Occup. Env. Hygiene.* 6(8): 665-671.

Mossman, J. L. and Sollito, M. A. 1991. Regulatory control of indoor radon. *Health Physics.* 60(2): 169-176.

# Section III

# THE MICROENVIRONMENT

# Chapter 8

# THE RESIDENTIAL ENVIRONMENT: HOUSING AND HEALTH

The home environment is considered by most people to be a safe haven and a refuge from the hazards encountered in the world at large. It is also the place where people spend the majority of their time, and for the elderly and infirm, almost all of their time. In this chapter I will review the myriad ways in which that home environment can pose hazards to one's health and some of the intractable societal problems involved in providing decent housing to the economically underprivileged among us.

## POVERTY AND HEALTH

It is not surprising that there is a strong correlation between poverty and ill health. Almost across the board, from infectious diseases to cancer, from heart disease to mental health, the poor are afflicted more frequently than the well-to-do. It is also no surprise that the same correlation applies to quality of housing. Our challenge is to separate out the effects of the physical environment from the many confounding factors which also contribute to poverty and ill health. Low educational level, poor diet, restricted access to medical care, the psychological stigma of poverty, and less incentive to care for rental housing as opposed to a home that one owns are all among those confounding factors. Substandard housing is clearly a factor in the potential health and safety problems which we will review. However, even newer, more expensive homes can have serious environmental defects—radon is one example. Another is mold contamination in wet wallboard caused by improper drainage design.

## ENVIRONMENTAL JUSTICE ISSUES

On a societal basis we have become very aware of the environmental justice issue. It is a fact of life in a market-driven economy that exposure to pollutants from industry, commerce, and the management of solid and hazardous waste will be greatest among the poorest segment of the population. By extension, it is perceived that such exposure contributes to the poor health status of that segment of society.

As environmental health professionals, we can only preach that proximity does not have to equate to hazard. Geographic separation, emission controls, and waste containment should be sufficient to assure protection for even the closest residences. Nonetheless, public perception remains that economic status dictates environmental injustices which must be redressed on a societal level. The problem is exemplified by attempts to locate nuclear or hazardous waste facilities on American Indian reservations in return for economic incentives.

## PEDIATRIC ENVIRONMENTAL HEALTH

It is generally believed that children are more susceptible than adults to the effects of pollutants. In some instances, such as exposure to environmental lead sources, that greater susceptibility has been well documented. The childhood lead problem will be explored further in this chapter. The need to protect children is particularly important in the residential environment because preschoolers spend so many hours in that environment and because a frighteningly high percentage of children in the United States live in poverty. While estimates of the percentages of children under 18 years of age living in poverty have tended to fluctuate somewhat, it is believed to be stubbornly greater than 20 percent overall and close to 50 percent for African-American children.

## HOMELESSNESS

We face a serious shortage of decent low-cost housing in the United States as we enter the twenty-first century. The limited supply of such dwelling units also results in serious overcrowding, which further contributes to health problems. However, the bottom end of the housing spectrum is the problem of no housing at all. Homelessness has been an issue in the United States since the 1980s. The extent of the problem is difficult to measure and fluctuates greatly geographically and seasonally. Estimates of the number of homeless people have ranged from 500,000 to several million. The homeless do not lend themselves easily to census counting. A major political issue in the late 1990s involved the concept of statistical sampling as opposed to actual headcounts during a census. It stems from the difficulty of finding and counting such groups as the homeless and illegal aliens. Needless to say, statistical sampling is embraced by liberals and vigorously opposed by conservatives, who fear that the increased counts will result in diversion of more federal dollars to those groups.

The homeless population is very diverse, ranging from those deliberately seeking a nomadic lifestyle to the chemically dependent and mentally ill who were at one time institutionalized. An increasing number are working families with small children who simply cannot afford unsubsidized apartments. Temporary shelters provide some relief from this problem but they too are often overcrowded. Poor health among this segment of the population is rampant, including high rates of infectious diseases, trauma, and chronic illness—all resulting from exposure, poor nutrition, and limited access to medical care.

## HEALTH-RELATED PROBLEMS OF SUBSTANDARD HOUSING

It is obvious that the relationship between health and housing is complicated by the entire gamut of socioeconomic issues besetting the poor—largely beyond the reach of environmental health professionals except in concert with other disciplines. In this section I will review some health-related housing issues which at least potentially lend themselves to solution through environmental intervention.

### Intestinal Disease and Outdoor Plumbing

One of the clearest associations between health and housing is that of intestinal disease rates among persons who do not have access to indoor toilet facilities. This is a rural rather than an urban problem but reminds us that substandard housing afflicts not only the inner city but rural areas as well. This is also a problem of economic status, because even in areas without municipal wastewater systems, individual septic tank and drainfield systems can suffice to prevent gastrointestinal illness from human wastes that contaminate water supplies or food.

### Vermin Infestations

Substandard dwellings invite serious infestations of rodents and insects. While some of these infestations are merely annoyances, others are directly associated with human health. One such example is the direct physical attack by domestic rats on vulnerable children and adults. In New York City alone more than 200 cases of rat bites are reported annually. In addition to serious physical trauma associated with such bites, a disease called *ratbite fever* can be caused by a specific bacterium (*Streptobacillus moniliformis*) transmitted through these attacks. Other examples include severe allergies to bedbugs, mites, and cockroaches. Environmental intervention that would prevent vermin infestation will be discussed in Chapter 15.

### Carbon Monoxide (CO) Poisoning

This issue, also discussed in Chapter 5, is applicable here as an example of a housing defect with serious health consequences. Improperly vented sidearm heaters are a very common defect which can result in CO release. Carbon monoxide detectors are now recommended for all dwelling units. Regular inspection of all gas appliances to detect defects is also recommended.

## Residential Fires

More than 3,000 people die in residential fires in a typical year, and more than 15,000 are injured—numbers which are significantly lower than 20 years ago. Residential fires are caused by numerous defects such as faulty wiring, wood-burning stoves, careless use of flammable solvents, and particularly by careless smoking habits (particularly among older adults). Fire prevention education and enforcement of electric wiring codes have contributed to reducing this problem.

The mandatory installation of smoke detectors has also been a major factor in reducing fatalities, if not preventing fires. However, even though a very high percentage of dwelling units have smoke detectors, about 50 percent of those devices become inoperable within a year of installation, primarily because batteries are not tested or replaced. Improving that record would undoubtedly prevent even more death and injury from residential fires.

## Temperature Extremes

One of the more insidious problems related to the residential environment is fatalities due to extreme heat (hyperthermia) or extreme cold (hypothermia). Almost every summer several hundred deaths (primarily of the very old) are attributed to extreme heat. In 1995 more than 500 people died in Chicago alone from this problem, and in 1998 about 100 more died during a Texas heat wave. In northern cities, apartment buildings are designed more to retain heat (for winter heating efficiency) than for cross-ventilation. In the Chicago deaths, other factors identified were failure to use fans or air conditioners because of concern over high utility bills, and keeping windows closed and locked because of fear of crime. Recognition of these issues has prompted many cities to initiate proactive plans for checking on the elderly during heat waves to assure that they are adequately cool.

Deaths from extreme cold often occur when heating systems break down in substandard dwellings. Similar proactive plans are needed to assure that isolated elderly citizens are cared for in those circumstances. Enforcement of building codes is also a necessary intervention.

## Childhood Lead Poisoning

Of all the issues associated with housing and health and the protection of young children, the effects of lead on children has received the most attention. It is somewhat ironic that much of this attention has come amid significant declines in blood lead levels over the last two decades of the twentieth century. Overall blood lead levels in the U.S. population declined 78 percent from the mid-1970s to the mid-1990s. At the same time, it has become evident that even low levels may cause subtle neurological changes and have a deleterious effect on learning ability and I.Q.

Table 8-1 summarizes the 1991 CDC recommendation for dealing with childhood lead.

Table 8-1

1991 CDC Guidelines on Childhood Lead Exposure

| Blood Lead Level (µg/DL) | CDC Recommendation |
|---|---|
| >10 | considered elevated |
| 20-44 | environmental abatement |
| 45-69 | treat by chelation |
| >70 | hospitalize (medical emergency) |

EPA data in 1990 indicated that some 3 million children in the United States under the age of six had elevated blood lead under that definition and some 250,000 had levels above 25µg/DL, which is considered the level of toxicity. Risk factors for elevation are clearly related to inner-city substandard housing and socioeconomic level. This knowledge has prompted CDC to back off from the recommendation that all children be screened and to suggest instead that only high-risk groups be targeted for screening.

*Sources of childhood lead exposure*

It is generally agreed that lead-based housepaints are the most important source. Prior to the 1940s, both interior and exterior housepaint was likely to contain up to 50 percent lead. Those levels were gradually reduced, but it was not until the 1970s that lead was entirely replaced by titanium as a pigment base. Thus, older housing stock, particularly in poor condition with peeling paint, remains an important source for young children. Window wells and door sashes, subject to frequent friction in use, are particularly likely to release lead. When children have *pica*—the desire to chew on unusual items such as paint chips and porch railings—the dangers increase. The discontinuation of lead-based paint and the gradual replacement of pre-1940s housing have undoubtedly contributed to the declining blood lead levels in the population at large.

A second recognized source of lead for children is soil, particularly around houses located in high-traffic neighborhoods. The lead originates from the leaded gasoline used before 1975 and only gradually reduced for some years thereafter until the number of cars using leaded gasoline declined significantly. Lead is persistent and relatively immobile in soil, thus requiring many years before it is significantly diminished. Pica again contributes to lead intake for young children with the condition; some actually ingest garden soil.

Other sources include lead from old water pipes (fortunately, the lead leaches out only when the water is acidic, which is relatively rare), soldered food cans (now almost completely replaced by lead-free seams), and ceramic glazing on cooking utensils. A few additional sources, such as imported slatted blinds and toys, have occasionally been identified. The phasing out of old housing (from approximately 30 percent to approximately 20 percent since the mid-1970s), disappearance of leaded gasoline, and the discontinuation of lead in food cans (from 47 percent to less than 1 percent) are cited as the major reasons for decreasing blood lead levels in the U.S. population.

*Environmental intervention for childhood lead exposure*

Although a medical treatment (chelation) exists to reduce moderately elevated blood lead levels, it is clearly preferable from a public health point of view to prevent elevation through environmental intervention. The difficulty is largely financial. Complete abatement of lead in a dwelling unit (stripping of all lead-containing paint from interior and exterior surfaces) is prohibitively expensive. Estimates are as high as $12,000 per dwelling unit in 1998 dollars. Furthermore, stripping can result in aerosolization, requiring occupational protection by respirators and evacuation of families during the abatement process. Abatement by paneling or wallpapering over the old paint is a temporary solution only. Stripping old paint from window wells and doorways is a less expensive, but only partial, solution.

*Federal jurisdiction*

One of the complicating factors in dealing with residential lead is the split jurisdiction among numerous federal agencies. While Health and Human Services (through CDC) has been responsible for surveillance and guidelines, the Environmental Protection Agency (EPA) has had jurisdiction over leaded gasoline; Housing and Urban Development (HUD) has been responsible for lead in housepaint; the Food and Drug Administration (FDA) has been in charge of cooking utensils and food cans; and the Consumer Products Safety Commission (CPC) has oversight of toys.

In 1996 a federal law (61FR9064-9088) was passed requiring disclosure of lead-based paints in real estate deals for all houses built before 1978. However, the law requires only disclosure and allows the buyer to seek further inspection. It does not require abatement.

## PREVENTION AND CURE FOR HOUSING DECLINE

Standards for housing quality were defined in the 1930s by an American Public Health Association (APHA) committee with reference to overcrowding (more than 1.5 persons per room), indoor plumbing, and physical dilapidation. The criteria included physiological needs (heat, light, ventilation), contagion (water, sewers, refrigeration), accident prevention (loose boards, wiring defects), and psychological provisions (privacy, normal family and community functions). The number of substandard dwelling units is, of course, a function of initial shoddy construction, outdated building codes, and aging. Maintenance and renovation of older housing units—or replacement with new units—are expensive and unevenly distributed. The problem extends to both rural and urban settings but has been most visible and pervasive in core cities.

The deterioration of inner city neighborhoods has been hastened by a host of social and economic circumstances which extend far beyond the reach of environmental health agencies. Although a few success stories may fuel occasional optimism, public policy has often been fragmented, ill conceived, and ineffective. It has also been fraught with political overtones and the classic dilemma of government versus private initiatives in search of solutions.

In this section I will review some of the approaches that have been tried—and the lessons learned from those that have not worked. Unfortunately, no panacea cure has been found, nor does one appear to be on the horizon. The same political, social, and economic obstacles remain, and the environmental interventions which are available to prevent illness and injury are often lost in the sea of poverty and despair. Too many Americans still live in substandard and dangerous dwellings, and the gulf between the haves and have-nots in society grows wider.

What can cities and metropolitan areas do to preserve housing stock and prevent blight? One essential approach is foresightful zoning and planning. Cities which have provided adequate greenspace (parks, plazas, etc.) and effective separation of industry, commerce, and residences have generally fared better than those which have not. Similarly, those communities which pursued building code updating and enforced codes most rigorously are better off than those which did not. In almost all American cities, the late twentieth century saw an acceleration of "white flight" to the suburbs in search of better schools and less crime. Inner-ring suburbs have experienced some of the same problems as the inner cities, subject to the same caveats of zoning, planning, and code enforcement.

## Historical Trends in Urban Renewal

The brief summary of federal programs and trends that follows reveals the ongoing frustration of urban blight. As the twentieth century ends, there is still a critical shortage of adequate housing for low-income Americans, and little inclination to expend significant money to provide such housing. Poverty (and substandard housing) remain concentrated in inner-city neighborhoods and attempts to disperse "housing projects" beyond the inner city are met with fierce opposition, resulting in ghettoization of the poor and all of their problems.

*Public housing projects*

As early as the 1930s, urban renewal projects were conceived to replace deteriorating neighborhoods with large high-rise public housing projects. The idea was that decent housing for the poor could be provided in these projects with new up-to-code construction and subsidized rents. These projects continued to be built in the post-war era but almost all have been colossal failures. Many have been demolished as their problems became intractable. What went wrong? Sociologists tell us that the projects attacked the symptoms of poverty but not the cause. Without addressing the economic and social problems associated with poverty, the projects simply concentrated the poor and their substance abuse, crime, and ill health in one place and accentuated the social ills accompanying poverty.

*The Model City approach*

In the 1960s, the Johnson administration presided over what some now call the "heyday of liberalism." There was optimism that government could solve the problem of poverty. The Model City approach was based on the lessons learned from failed public housing projects. The causes would be attacked along with the symptoms. Areas for renewal were designated and the process separated from entrenched city hall politicians. The people affected would decide for themselves how to spend the money allocated. Model police precincts, health clinics, job opportunities would be funded, and renovation of old houses would be part of the mix rather than new high rises.

Did that approach work? Clearly, the problems of poverty and inner-city decay were not solved. To conservatives it became a symbol of big government folly and failure of the liberal approach. To liberals it was a case of a good idea not given sufficient chance to work. They claimed that the Vietnam War, federal deficits, and inflation intervened to end the experiment prematurely. The election of the Nixon administration in 1968 signaled a new, more conservative era with different ideas.

*Block Grants*

Under the Nixon administration, Block Grants to cities became the funding mechanism for urban renewal. The decision-making process was returned to City Hall, and the worst inner-city neighborhoods again had to compete for attention with the many other urban needs. Elements of the diversified approach and the recognized need to attack causes rather than symptoms survived and remain part of the mix in the 1990s.

*Dispersal and homesteading*

One of the lessons learned from earlier failures was that smaller, more dispersed low-rise projects had a greater chance of success than concentrated high-rises. In addition, individual dwellings, wherever located, could be acquired by the city and sold to individuals for a nominal fee on the condition that they be renovated and brought up to code through the mechanism of low-interest loans (homesteading). Mixed-income housing projects (some subsidized and some market-value rental units) are also part of that concept. These approaches have had a much higher success rate. They are subject, however, to the ongoing opposition to the spread of low-cost housing into established higher income neighborhoods. The resistance is fueled by the fear of introducing higher crime rates, future neighborhood deterioration, and ultimately, reduced property values. If this opposition is ever to be overcome, the fears will have to be proven unfounded and such projects shown to work to the advantage of all concerned.

*Gentrification*

In many cities, one or more inner-city neighborhoods has become gentrified. This phenomenon started in the 1970s with the realization that there were advantages to being close to downtown. Homes were extensively renovated, and shopping streets with restaurants, boutiques, and fancy lighting were created. The gentrified neighborhoods attracted the well-to-do and were unquestionably upgraded, providing tax revenue for the city. However, this gentrification provided few benefits for the poor. They were simply displaced as the old neighborhoods became unaffordable.

*Enterprise Zones*

The Bush administration (following the Reagan years, which provided no significant initiatives in low-cost housing) fostered the concept of "enterprise zones" in the inner city. The concept was that private industry could do a better job than the federal government of solving the problem of housing for the poor. The government would provide tax incentives for capitalists to build reasonable housing profitably. After being promoted heavily by Jack Kemp, former secretary of HUD, the concept lost steam, and followup results were not encouraging. The Clinton administration in the 1990s promoted the idea of a public/private partnership for inner cities, but the budget balancing emphasis has not provided significant funds or incentives in this area.

*Welfare Reform*

The much-heralded Welfare Reform Act of 1997 could radically alter the future of public involvement in the plight of the poor and ill-housed segment of society. What effect that reform may have on low-cost housing remains to be seen as this is written. Reducing the welfare rolls and replacing welfare with workfare will not automatically increase the supply of affordable housing. While the welfare roles may be reduced, I believe that affordable housing for low-income people will still have to be subsidized in some manner if the problem is to be solved.

*Brownfields Initiative*

The late 1990s saw the advent of still another buzzword for urban renewal. The "Brownfields" initiative refers to the revitalization of abandoned and decaying commercial and industrial sites, often on hazardous waste–contaminated land. Cities are often punctuated with such sites, ranging from abandoned corner gas stations to large industrial sites in "rustbelt" communities. Redevelopment usually includes expensive "superfund" cleanup, but a start must be made on reclaiming such properties for a variety of industrial or public uses. The term LULUs (locally unwanted land uses) has been added to the buzzword brigade, referring to undesirable, derelict neighborhood eyesores such as junkyards or hazardous waste treatment facilities. When Brownfield sites are redeveloped, residents are asking for input into the process to assure that additional LULUs are not created.

*Turn-of-the-Century Trends*

As we enter the next millennium, there are some interesting trends in how Americans are housed. At one end of the spectrum is the trend toward highly secure gated communities, where the well-to-do isolate themselves securely from the rest of society. The opposite trend, occurring in many cities, is sometimes referred to as the "new urbanization." It is a return to the old village square concept of an open community with front porches, sidewalks, and storefronts accessible for all without reliance on cars. This concept decreases isolation and is even being tried in suburban and exurban communities. Such developments, however, are also mostly unaffordable for those most in need of low-cost housing. Where these trends will lead us in the twenty-first century is anybody's guess. The need remains critical to find a way to provide decent housing for low-income Americans, and our free-market economy does not seem to abound with viable solutions.

**REFERENCES**

References related to minimum housing standards, the CDC statement on childhood lead poisoning and new trends in urban redevelopment are as follows:

Mood, E. (Ed.) 1986. *Housing and Health: APHA-CDC Recommended Minimum Housing Standards*. American Public Health Association. Washington, D.C.

CDC, 1991. "Preventing Lead Poisoning in Young Children." A statement by the CDC. William L. Roper, M.D., Director.

Greenburg, M., C. Lee and C. Powers. 1998. "Public Health and Brownfields: Reviving the Past to Protect the Future." Editorial. *AJPH* 88(12): 1759-1760.

# Chapter 9

# THE WORK ENVIRONMENT: OCCUPATIONAL HEALTH

Second only to the home, the work environment is where most Americans spend a significant portion of their total hours. The variety of activities and potential exposure to hazardous physical, chemical, and biological agents is so extensive that the occupational environment is often considered separately from other environmental health categories (e.g., departments of Environmental *and* Occupational Health). It also involves a variety of subdisciplines including Occupational Medicine, Occupational Health Nursing, Industrial Hygiene, and Occupational Injury Control.

Occupational health is dominated nationally by the Occupational Safety and Health Act (OSHA) of 1970 and the agency created for its enforcement, the Occupational Safety and Health Administration (also OSHA), administered through the Department of Labor. OSHA covers all private-sector employees except the self-employed and family farms. Many government employees are also covered through executive order or state plans. The Act includes a General Duty Clause stating that "each employer shall furnish a place of employment free from recognized hazards which are likely to cause death or serious physical harm." The clause applies to all covered workplaces and specifies basic safety features such as sufficient restrooms, guarded machinery, and adequate ventilation. Specific standards have periodically been promulgated spelling out detailed requirements for more dangerous occupations. However, debate over the effectiveness of OSHA continues. Conservatives in Congress see it as a sop to labor and an excessive burden on industry, while liberals see it as a vital protection for the rights of workers.

In addition the Act created the National Institute for Occupational Health and Safety (NIOSH), placed administratively in the Centers for Disease Control and Prevention (CDC). NIOSH is essentially the research and education arm of OSHA.

## SCOPE OF OCCUPATIONAL HEALTH PROBLEMS

The record-keeping requirements of OSHA provide an ongoing accounting of annual injuries, illnesses, and fatalities related to the workplace. The required reports include fairly accurate numbers for injuries and acute illnesses, but they are less able to define the role of the occupational environment in chronic disease causation because of the many confounding factors outside of the workplace.

Better record keeping and reporting may result in higher numbers even when true incidence is declining. The numbers fluctuate somewhat from year to year but do provide a picture of the hazards of the workplace in the late 1990s. They tell us that the incidence of illness and injury in the workplace is approximately eight per 100 full-time workers. This translates to more than 6 million lost-time injuries and 500,000 illnesses in a typical year at a cost estimated to be about 120 billion dollars. Clearly, the hazards are much greater in some occupations than in others. Construction workers, farmers, miners, truck drivers, and nursing personnel are among the high-incidence workers, while white collar jobs are much safer.

Mortality data are also very revealing. There are more than 6,000 occupational fatalities annually, a full one-sixth of those among construction workers. This translates to a rate of approximately six per 100,000. Some experts believe that the true rate is much higher because long-term chronic effects of occupational exposure as a contributor to mortality are not included. By one estimate, there may be as many as 50,000 additional fatalities each year for which chronic occupational exposure is a major contributing factor.

The major cause of occupational fatality is motor vehicle accidents, accounting for about 20 percent. In second place is homicides with about 16 percent or more than 1,000 deaths each year. For women, who are less likely to be employed in the most hazardous job categories, homicides account for almost 40 percent of the fatalities. Many of those involve retail clerks, particularly those working night shifts in convenience stores or service stations. The magnitude of that problem has spurred efforts to institute protective measures such as not working alone, secure cashier stations, and alarm systems. Violence in the workplace has increasingly been recognized as a major occupational health problem spurring increased efforts to understand causes and to prevent incidents. Other significant causes of occupational fatalities include falls (approximately 10 percent) and being struck by an object (approximately 9 percent).

## INDUSTRIAL HYGIENE MEASUREMENTS

Industrial hygienists are concerned with occupational illnesses caused by chemical, physical or biological agents. They are concerned with both acute effects (trauma, toxicity, and infection) and with longer-term chronic effects of those agents, including cancer and such specific chronic conditions as asbestosis and silicosis. Chronic effects are more difficult to ascertain because they are often confounded by nonoccupational exposures such as diet, tobacco use, and community air, water, and food contaminants. However, personal and area exposure measurements in the work environment provide data which can be useful in epidemiological investigations of occupational illness.

There are numerous examples of definitive association of occupational exposure with chronic illness from such epidemiological studies. Scrotal cancer in chimney sweeps, leukemia in radium dial painters, mesothelioma in asbestos workers and lung cancer in uranium miners are all examples of chronic disease associations.

Some of the measurements which are used to ascertain exposure and determine compliance with standards are as follows:

### Time Weighted Average

Time-weighted average (TWA) is the actual measurement of worker exposure to a particular substance, usually based on an eight-hour work shift. It is used to determine compliance with OSHA regulatory standards. The measurements are averaged over time to even out peaks and valleys inherent to many job tasks.

### Threshold Limit Value

Threshold limit value (TLV) is a consensus standard developed by the American Conference of Government Industrial Hygienists (ACGIH) from known data sources (published research) to describe the TWA below which ongoing eight-hour workshift exposure will not result in harm for almost all workers. The ACGIH publishes an annually updated list of TLVs for some 850 substances (expressed in ppm or $mg/m^3$). The TLV is not an enforceable standard but provides a scientific basis for determining potential health effects.

### Permissible Exposure Limit

Permissible exposure limit (PEL) is the enforceable regulatory limit set by OSHA, often based on the published TLV for a given substance.

### Action Level

The action level (AL) is also an OSHA mandated TWA standard. The AL is set below the PEL to trigger corrective action when the concentration of a given substance approaches but does not exceed the PEL.

### Short-term Exposure Limit

Another OSHA standard, the short-term exposure limit (STEL) is usually based on a 15-minute period. It is used to assure that workers are not exposed to dangerous levels of a substance for short periods even when the PEL is not exceeded when averaged over the eight-hour workshift.

### Ceiling Value

The ceiling value (CV) is the peak value mandated by OSHA, which should not be exceeded even instantaneously regardless of TWA exposures over longer periods.

### Recommended Exposure Level

The recommended exposure level, or REL, is a NIOSH recommendation, not binding on the enforcement agency.

### Occupational Exposure Limit

The occupational exposure limit (OEL) is an industry-generated recommendation without regulatory implications.

### Biological Exposure Index

BEIs are nonenforceable limits recommended by ACGIH for substances measured in blood or urine as opposed to airborne area or personal exposure measurements.

## CHEMICAL EXPOSURES

There are countless potential chemical exposures that may be encountered in industrial processes. Exposure may be via inhalation or skin absorption or occasionally by ingestion. Acute exposures include chemicals which are explosive, corrosive (acids and bases), irritating, flammable (flashpoint of 60°C or less), oxidizing (oxygen release causes fire or violent reaction with water), or toxic. This list understandably parallels the definitions for hazardous waste discussed in Chapter 6. Chemical exposures of concern for potential chronic effects include the familiar carcinogenic, teratogenic, and mutagenic categories.

Some examples of chemical exposures of particular concern to industrial hygienists include:

### Solvents

Solvents such as benzene, toluene, alcohol, and turpentine are dangerous because they may be either inhaled or absorbed through the skin.

### Particulates

Particulate aerosols of particular concern are respirable particles (of less than 5μm aerodynamic diameter) such as asbestos, silicon dust, or cotton dust which may directly affect the lungs.

### Irritants and Sensitizers

Primary irritants and chemical sensitizers (skin or lungs) include formaldehyde, epoxy resins, and organic solvents.

### Heavy Metals

Common heavy metals include lead, mercury, or cadmium. In the workplace a blood lead level greater than 25μg/DL is considered elevated and greater than 60μg/DL requires removal from the workplace. These levels are higher than the childhood lead levels discussed in Chapter 8 because adults are less susceptible to neurological impairment from lead than are young children.

## PHYSICAL EXPOSURES

Physical agents of potential health concern in occupational settings include such things as noise, excessive heat, and radiation.

### Noise

Noise is defined as any unwanted sound. Loud rock 'n roll music is an example of a sound which may be considered music by some but noise by others. Noise is measured in decibels (sound pressure on a log scale) and in frequency (hertz or cycles/second). Excessive exposure may have serious physiological effects ranging from temporary nausea to permanent hearing loss. It can also interfere with communication, thus creating an additional safety hazard. Table 9-1 lists decibel levels associated with particular activities.

Table 9-1

Activities Associated with Various Decibel Levels

| Decibel Level | Associated Activity |
|---|---|
| 50 | Average office background |
| 60 | Normal conversation (3 ft.) |
| 80 | Noisy bar or restaurant |
| 85 | 8 hr TWA -OSHA AL |
| 90 | Subway- Also 8-hour TWA OSHA PEL |
| 100 | Textile mill |
| 110 | Symphony orchestra or chainsaw; Also, 15-minute OSHA STEL |
| 115 | OSHA ceiling value |
| 120 | Hydraulic press |
| 140 | Jet engine |
| 180 | Rocket launching pad |

## Heat

Excessive heat is another physical exposure of concern. There are three types of effects from heat exposure:

### *Hyperthermia*

Hyperthermia (or heat stroke) from prolonged exposure to high temperatures. The body is unable to maintain cooling, heat regulation breaks down, the face becomes flushed, and death can ensue. Prevention is primarily achieved by worker rotation to avoid prolonged exposure. Hyperthermia was described as a problem for the elderly in Chapter 8.

### *Heat cramp*

Heat cramp results from heavy exertion in hot environments. It is marked by loss of salt and moisture and requires salt tablets and increased water intake for prevention.

### *Heat exhaustion*

Heat exhaustion is also the result of excess physical exertion and is marked by a decrease in body temperature.

### Radiation

Radiation as a physical occupational hazard was discussed at some length in Chapter 7. Such well documented examples as leukemia in radium watch dial painters and lung cancer in uranium miners have provided evidence supporting the protective intervention strategies since promulgated.

## BIOLOGICAL EXPOSURES

In addition to chemical and physical exposures, some occupations also present the threat of exposure to infectious biological agents. Healthcare workers, laboratory workers, meat packers, pharmaceutical workers, and many segments of the agricultural industry all face biological hazard potential. The highly publicized AIDS epidemic starting in the 1980s raised additional concerns about occupational blood-borne disease (including AIDS and hepatitis B) extending to law enforcement personnel, prison guards, and embalmers.

More than 120 cases of occupationally acquired AIDS have been reported by CDC. In response OSHA developed a specific Blood-borne Disease Standard, which became effective in March 1992. The reemergence of tuberculosis in the late 1980s and early 1990s further accentuated the potential for occupationally acquired infections. In 1994, OSHA implemented a tuberculosis standard requiring compliance with established CDC guidelines pending development of a specific OSHA tuberculosis standard.

## ERGONOMICS

Occupational hazards related to body mechanics, posture, eyestrain, and work stress are grouped under the heading of *ergonomics*. It is estimated that about 20 percent of occupational injury and illness involves ergonomically related back injury alone. Considerable progress has been made in designing ergonomically sound tools and workstations. With an increasing number of workers spending a high proportion of the workday sitting in front of computers and word processors, ergonomic design of such workstations to reduce backstrain and eyestrain has taken on considerable importance.

Cumulative Trauma Disorders (CTDs) related to ergonomic deficiencies have been a very rapidly increasing phenomenon, with more than 300,000 cases occurring annually in the 1990s. Carpal tunnel syndrome is the most common of these disorders. It results from repetitive motion of the hand and wrist, sometimes requiring surgery and leading to prolonged disability. It is particularly common among supermarket cashiers and meat and poultry packers. Interventions include better tool design and worker rotation.

Raynauld's disease, a CTD affecting the fingers and toes, arises from persistent vibration activities such as jackhammer operation.

## CHRONIC DISEASES

Although long-term effects of trace contaminants are difficult to assess occupationally just as they are in the community, the unique nature of some industrial activities has provided more definitive evidence of occupational hazards causing chronic disease. Several radiation induced cancer examples have previously been mentioned.

Asbestos is another example for which there is a clear occupational association with three specific long-term chronic effects. These include asbestosis, lung cancer, and mesothelioma (a rare, highly fatal cancer of the abdominal lining highly correlated with asbestos exposure). Asbestosis and lung cancer risks are greatly multiplied by concomitant smoking habits, but all of these conditions are associated with occupational exposure to asbestos in miners, shipbuilders and other asbestos-related industries. There is a 20- to 25-year latency period and many of the reported cases are in individuals whose occupational exposure predated the OSHA regulations. The 1998 OSHA standard is 0.1 fibers/$cm^3$.

Asbestos was used extensively in building materials from the 1940s through the 1970s for its excellent fire retardant qualities, before its hazards were recognized. Much of the current occupational exposure results from asbestos abatement activities. Regulations require asbestos removal before demolition of asbestos containing buildings. They also require asbestos removal or containment before renovation projects that may aerosolize asbestos fibers. Elaborate isolation barriers and personal protection equipment are routinely used for such activities.

Considerable controversy has raged over the potential for public exposure to friable (flaking) asbestos in buildings, which is also released from brake linings into ambient air. The controversy particularly affected public schools, which were required by federal law enacted in 1980 to identify and abate asbestos at a cost often running to hundreds of thousands of dollars per school. Many scientists questioned the reasoning that asbestos posed a significant threat to children based on extrapolations from past occupational exposure. Extrapolating from high-level occupational exposures to low-level public exposures is very uncertain, particularly in light of the different hazards posed by different types of asbestos. About 98 percent of the asbestos incorporated into the building materials was *chrysotile* (a curly type of fiber) which is perceived to be less hazardous than *amphibole* (a straight, pointed fiber), which made up the other 2 percent.

## INDUSTRIAL HYGIENE INTERVENTION

The intervention strategies promulgated for preventing occupationally acquired illness or injury are predicated on the "industrial hygiene hierarchy." The concept is that all job tasks should be designed to be safe, using engineering and administrative mechanisms rather than relying on individual worker action. The hierarchy includes the following elements:

### Industrial Hygiene Hierarchy

The accepted industrial hygiene hierarchy consists of four elements:

1) The level of protection must be reliable, consistent, and adequate.

2) Efficacy of the protection must be determinable for each worker.

3) Dependence on human intervention must be minimized .

4) All routes of exposure must be considered.

*Engineering controls*

The first priority is engineering controls that minimize the need for individual preventive actions. Primary examples of these controls include the following:

1) Substitution of materials—using a less harmful substance whenever technically feasible, such as substituting water for a chemical solvent.

2) Change to a safer process—such as converting to an airless paint sprayer to reduce the aerosol release.

3) Enclosure or isolation of a process—such as using lead shielding for radiation, doing manipulations within glove boxes, or adding sound muffling to machinery.

4) Dust control by hosing down a coal pile.

5) Installing local exhaust fans over a particular operation, such as exhaust formaldehyde over an autopsy table.

6) Installing general exhaust fans to provide adequate dilution.

7) Encapsulation with hardening materials—applicable to asbestos-lined pipes or ducts to prevent flaking of loose (friable) material.

8) Continuous monitoring—including preset alarms and automatic shutoff when concentrations approach permissible limits.

*Administrative controls*

The second priority is administrative controls, which include:

1) Rotation of workers. This concept has been mentioned as applicable to extreme temperature environments or repetitive motion tasks where neither the environment nor task can be modified, but exposure time can be limited to prevent the problem.

2) Training. The Federal Hazard Communication Standard mandates that workers be informed of potential workplace hazards and adequately trained to perform duties safely before being assigned to those duties. Many states have also passed right-to-know legislation. Periodic updates of this training are also required. It has been well documented that the safety climate in a workplace, defined by the degree of officially declared management commitment to worker safety, is a major factor in occupational health.

3) General cleanliness and housekeeping can also be included under the heading of administrative actions.

*Personal protective actions and devices*

The lowest priority in the industrial hygiene hierarchy is personal protective devices. Such devices are considered to have the greatest potential for failure. However, in certain situations they become a vital element in the safety program. Examples include:

1) Personal respirators. Respirators are used when ventilation controls cannot assure maintaining the concentration of a substance below the PEL. Suitably chosen and fitted respirators can maintain concentration of many substances below the PEL when ventilation systems are not able to do so.

2) Earplugs. These provide noise control in situations where source muffling is insufficient (such as for jet airplane mechanics).

3) Safety glasses, safety shoes, and other protective clothing directly shield vulnerable parts of the body.

4) Vaccination may be a less obvious, but higher priority, form of worker protection. Rabies vaccination for veterinary workers and hepatitis B vaccination for some healthcare workers are examples of well accepted and effective personal protective actions.

*Medical surveillance*

Another approach to worker protection which is somewhat controversial is medical surveillance. Preemployment physicals may be useful for initial job placement. At that time serum samples can be collected which may be useful for determining preexposure, for instance, to a pathogen or to heavy metals. The value of such information relative to cost is still debated. It is generally agreed that immediate reporting of known exposures for immediate treatment and medical followup is a worthwhile feature of an employee health program. Proactive encouragement of healthy life styles through employer-provided smoking cessation programs, exercise programs, and workout facilities is also practiced by some organizations as a means of keeping workers fit and potentially reducing health insurance and/or workers' compensation costs.

## REFERENCES

A very useful textbook on the subject of occupational health is:

Levy, B.S. and D.H. Wegman, 1995. *Occupational Health: Recognizing and Preventing Work Related Disease.* (3rd Ed.) Little Brown. Boston, MA.

# Chapter 10

# THE INSTITUTIONAL ENVIRONMENT: NOSOCOMIAL INFECTIONS

The last of the microenvironments is that of institutions. An institution is a self-contained property with a defined (and often very susceptible) population. Some primary examples are healthcare facilities, daycare centers, schools and colleges, prisons, military installations, resorts, and migrant labor camps. Clearly, these are very diverse settings with extremely different populations at risk.

In occupational settings, the population is assumed to be primarily healthy adults. By contrast, institutional populations often comprise uniquely vulnerable individuals. In daycare centers, they are young children, often untrained in basic sanitation, with a high potential for transmitting enteric and respiratory illnesses to each other and their caregivers. In healthcare facilities, they are increasingly the very ill and immune-compromised segment of the population requiring unique protection from the environment. In institutions of higher education, the population is primarily young adults, and environmental protection centers on the unique array of hazardous physical, chemical, and biological agents abounding in campus research laboratories. Prison populations are also mainly young adults, but with a high degree of chemical dependency and psychiatric problems that make them uniquely susceptible to psychiatric stress. Prison facilities also present unique problems in fire safety, where the usual emphasis on accessible emergency exits is not appropriate. Finally, military recruits have been found to be susceptible to respiratory infections related to stress and crowded living conditions.

Common to all of these situations is that environmental protection must be specifically designed for the populations at risk, often encompassing interventions not applicable to community or occupational environments. In this chapter I will use the healthcare environment as the primary example of such approaches, concentrating on the example of nosocomial (hospital-acquired) infections to illustrate the need for environmental protection over and above standard community interventions.

## SCOPE OF NOSOCOMIAL INFECTIONS

Changing healthcare patterns in recent years have had a profound impact on hospitals and clinics. The trend has been a decline in the number of hospitals, from about 7,000 to about 6,300, and in the number of beds, from 6.4 million to about 6.1 million, since 1984. Shorter stays for many types of patients and greater utilization of outpatient clinics even for surgery have fueled this decline.

As a result, patients who are hospitalized are generally the sickest and least resistant to infection. Thus, despite continuing infection control research and effort, the overall rate of nosocomial infections per 100 discharges has remained relatively constant at approximately 5 percent. This translates to some 2 million infections annually, resulting in increased hospital stays averaging more than four days at a cost now exceeding $1,000 per day. In 1992, nosocomial infections caused directly approximately 19,000 deaths and contributed to 58,000 others. The totals indicate a major problem relative to the scope of other infectious diseases in the United States today. However, from an environmental health perspective it is important to note that only some of these illnesses are preventable. The greatest single factor is host susceptibility, the link in the epidemiological chain least amenable to environmental intervention.

Data on nosocomial infections are also difficult to obtain. Hospitals are understandably reluctant to advertise such statistics. Our knowledge comes from a CDC effort known as the National Nosocomial Infections Surveillance Study (NNISS). About 125 sentinel hospitals are recruited for this study to utilize standardized criteria to report their infections to CDC. National data are extrapolated from these sentinel hospitals. Another CDC study, the Study on the Efficacy of Nosocomial Infection Control (SENIC), found that only 25 to 30 percent of reported infections could be considered preventable through known infection control practice, again due to the very important host susceptibility factor. However, it was also revealed that only about a 10 percent reduction was being realized, leaving room for considerable advances. The main incentive for hospitals to prevent these infections is the concept of Diagnosis Related Groups (DRGs), whereby reimbursement is based on a patient's diagnosis, not the number of days the patient is hospitalized.

The rise of Health Maintenance Organizations (HMOs) is also having a considerable impact. The shorter stays make it more difficult to track infections, but they also reduce exposure of patients to the healthcare environment.

## HISTORICAL NOTES

While the role of the environment in nosocomial infections has been the subject of debate and disagreement, it is clear that hospitals have come a long way in providing a cleaner, safer environment since the nineteenth century. Before the microbiology era was ushered in by the discoveries of Louis Pasteur and Robert Koch, among others in the second half of the nineteenth century, hospitals were known as "pest houses." They were places where people went to die. Sanitary conditions were appalling, and, in the absence of comparable records, we can assume that almost all patients acquired infections from that environment.

A few early reformers stood out for their observations of these conditions even before the germ theory of disease was proven. Florence Nightingale is known as the mother of modern nursing for her observations and efforts in a British military hospital during the Crimean War in the 1850s. She advocated cleanliness and sunlight and demonstrated a major reduction in mortality among wounded soldiers when her ideas of clean bedding, frequent dressing changes, and scrubbing of the environment were implemented.

Ignaz Semmelweiss was a physician in the Lying-In (maternity) hospital in Vienna in the 1840s. He observed that the rates of puerperal fever (a highly fatal streptococcal infection in postpartum women) were much higher in the section of the hospital overseen by physicians and their medical students than it was in the section presided over by midwives. He postulated that lack of handwashing by the physicians even after performing autopsies on puerperal fever victims was a major cause of these infections. He instituted a mandatory handwash with chloride of lime and demonstrated a major resulting drop in puerperal fever rates. His work remains today as the major justification for emphasis on handwashing in healthcare facilities.

Joseph Lister was a Scottish surgeon in Glasgow in the 1860s, a time when postsurgical infections were rampant. He believed that most of these infections were preventable and used a strong phenol solution to soak instruments before use. He also sprayed the phenol into the air (a practice which would now be frowned on by OSHA) and demonstrated that infection rates could be reduced.

Once the germ theory of disease had been conclusively demonstrated, hospitals rapidly changed character and the "age of antisepsis" was ushered in. Steam autoclaves were introduced as early as the 1890s to sterilize instruments, and the environment was subjected to disinfectant scrubs. "Hospital Clean" became the hallmark of patient care, and, again without comparable records, we can assume that nosocomial infection rates were reduced to levels approximating today's.

In the 1940s, antibiotics were introduced that could actually cure bacterial infections. However, that development may also have generated an overconfidence that led to a letdown in aseptic technique. Staphylococcal infections cropped up in newborn nurseries and surgical suites, fueled partially, early evidence suggests, by antibiotic resistance. This led to renewed interest in the role of the environment in these infections. But antibiotics also led to an accelerating trend toward more chronic disease patients and fewer infectious disease patients. Other advances in medicine, including cancer chemotherapy, radiation treatments, and organ transplantation, were creating a chronic disease patient population more susceptible to nosocomial infections.

This soon turned into what has become known as the Opportunist Era. Opportunists are microorganisms which are not normally considered pathogenic to healthy humans but cause infection in individuals whose immune mechanisms have been compromised by disease or treatment. Cancer patients, diabetics, organ transplant patients, and AIDS patients are all in that category. The opportunists are primarily gram-negative bacteria, including common intestinal organisms such as *Escherichia coli*, *Pseudomonas,* and *Klebsiella* species. Many of these opportunists were resistant to the early antibiotics, enhancing their competitive position in the hospital environment. Other organisms, such as the protozoan *Pneumocystis carinii*, which has significantly affected AIDS patients, have since been added to the list.

In the 1990s, we saw the emergence of many new pathogens and the reemergence of many old ones. Some of these are important in nosocomial infection. Particularly important have been the antibiotic-resistant strains, which now tax our capacity to keep up with weapons to fight infection. Methicillin- and oxycillin-resistant staphylococci are prominent in this category, but the emergence of vancomycin-resistant enterococci is particularly perplexing. Vancomycin has been the antibiotic of last resort as staphylococci and other organisms have become more antibiotic resistant.

Emerging signs that vancomycin resistance may be transferred to these organisms raises the specter of reduced ability to cope with bacterial infections, potentially returning us to the pre-antibiotic era of medicine. The percentage of vancomycin-resistant enterococcal isolates in U.S. hospitals increased from 7.1 percent in 1992 to 16.7 percent in 1996. In an effort to combat this trend, many hospitals are restricting vancomycin use and returning to basic principles of environmental control and handwashing, while continuing to search for advanced pharmaceutical solutions.

## THE EPIDEMIOLOGIC CHAIN

Nosocomial infections are characterized by unusual circumstances affecting all of the links in the epidemiologic chain. Some of those unique features are as follows:

### Agent

As mentioned, many of the agents causing nosocomial infections are "opportunists," which differ from other pathogens in that they do not have inherent capacity to breach cell barriers unless that barrier is weakened by a host condition such as immunosuppression. The increasing problem of antibiotic resistance is also accentuated in nosocomial isolates.

### Reservoir

The healthcare facility is marked by the unusual circumstance of combining the most susceptible hosts (traumatized and immunocompromised patients) with the most infectious human reservoirs (infectious patients). In addition, there is the unique problem of susceptible hosts serving as their own reservoirs. Many nosocomial infections are "autogenous" meaning that the victim acquires the agent from a different colonized site on his or her own body. Many urinary tract infections, for example, are with intestinal *E. coli* which migrate to the urinary tract through catheterization of the patient. Surgical wounds likewise are often infected with opportunists from the skin or intestinal tract of the victim.

### Escape/Release

Escape of organisms from hospital patients is exacerbated by the high degree of incontinence, sneezing, coughing, and wound exudates associated with this population.

### Transmission

The unique avenue of autogenous transmission is particularly difficult to intercept through environmental intervention. When patients are severely immunosuppressed, organisms which migrate from a colonized body site to a site unprepared for a microbial challenge are highly likely to cause infection. (A colonized site is one in which normal flora reside commensally and without disease implications on or in the human body.) The various environmental intervention strategies designed to interrupt pathways of transmission (barriers or cidal treatments) are largely ineffective against autogenous transmission. Efforts are therefore directed principally against the exogenous mechanisms of transmission. Those include staff-to-patient or patient-to-patient direct transmission and environmental avenues such as air, water, or medical instruments or devices. Environmental intervention strategies will be discussed later in this chapter.

### Portals of Entry

Unusual pathways to the new host are opened up by various medical interventions. They include surgical incisions into normally sterile body regions, intravenous hookups for fluids or medications, urinary catheterization, and respiratory assistance devices often combined with water reservoirs for humidification.

### Host Factors

The various degrees of debilitation, immunosuppression, and trauma which afflict hospitalized patients are the major factors in nosocomial infections and are responsible for the low percentage of preventable infections amenable to environmental intervention.

## TYPES AND DISTRIBUTION OF NOSOCOMIAL INFECTIONS

While the distribution of the various types of nosocomial infection varies considerably from hospital to hospital and with time in the same hospital, the approximate generic breakdown is summarized in Table 10-1.

Table 10-1

Types and Percentage of Nosocomial Infections

| Body Site | Percentage |
|---|---|
| Urinary tract (UTI) | 40% |
| Respiratory | 20% |
| Surgical site | 20% |
| Skin | 5% |
| Bacteremia | 5% |
| Gastrointestinal (GI) | <3% |

Bacteremias (or septicemias) are generalized infections that may result from intravenous catheters or from systemic problems resulting from extreme debilitation. They are the infections with the highest fatality rates. The low percentage of GI infections stems from the inherent design of the gastrointestinal tract to accept microbial challenges. GI infections still pose a threat in some situations, such as for newborn and premature infants.

The distribution of agents responsible for nosocomial infections also varies greatly from hospital to hospital. The leading causative agent for many years has been the familiar *E. coli.* That agent is responsible for about 20 percent of all nosocomial infections and more than 30 percent of the UTIs. Other prominent agents include the ubiquitous *Staphylococcus aureus* and the *enterococci,* which are now becoming resistant to vancomycin. They each cause about 10 percent of reported infections. Other gram-negative enteric bacteria in the opportunist group that are commonly reported include *Klebsiella, Pseudomonas,* and *Proteus.* Another group is the opportunistic fungi. Among cancer patients, *Candida albicans* is the most prominent, while *Aspergillus fumigatus* is a dangerous pathogen for bone marrow transplant patients. The protozoa *Pneumocystis carinii* and *Cryptococcus* are of particular danger to AIDS patients, both in the hospital and in the community.

## INTERVENTION STRATEGIES

We have seen that prevention strategies to minimize the nosocomial infection problem are generally limited to those transmitted exogenously. It is also possible that environmental control can limit new colonization once a patient is hospitalized and that newly colonized organisms are more likely to cause autogenous infections than are previously established colonizers. The role of the environment in these infections has remained controversial, but general practices to limit contamination are usually considered aesthetically defensible, even without epidemiological evidence of disease prevention.

Prevention can be looked upon as a classic contamination control problem, which, in the healthcare environment, must include elements of both containment and exclusion. *Containment* is using barriers to contain a hazardous agent known to be present (e.g., a patient with a diagnosed infectious disease) and prevent its escape. *Exclusion* is the opposite concept, where a patient is known to be ultrasusceptible to infection, and barriers are used to keep out any potential infecting agents. The primary example of the exclusion practice in healthcare facilities has been the well-established surgical aseptic practice utilizing hand scrubs, gloving, gowning, and masking to prevent contamination of the surgical site.

Accredited hospitals must have organized infection control committees to oversee mandatory written policies for the surveillance and prevention of nosocomial infections. These committees follow published CDC guidelines as well as guidelines promulgated by professional infection control organizations. State regulations also prevail in many areas, and occupationally, OSHA standards for blood-borne diseases and tuberculosis, discussed in Chapter 9, must be adhered to.

### The Historical Isolation Concept

Until the 1990s, conventional isolation practice in healthcare facilities was diagnosis driven. Patients were assigned to various categories of infectious isolation (containment) on the basis of their diagnosis. The categories included respiratory, enteric, wound and skin, and a strict isolation category for diseases with multiple means of spread. There were also specific categories for diagnoses such as tuberculosis, hepatitis B, and AIDS. In addition, until 1983 there was a specific protective isolation category (exclusion) for immune-compromised patients. That category was dropped when it was concluded that strict adherence to the restrictive procedures might compromise patient care, and more flexibility was needed. However, the principles of protection encompassed in that category are still very much in evidence in surgical suites and in nursing units housing particularly susceptible patients.

### Standard Precautions

In the late 1980s, in response to concerns with occupational blood-borne disease (AIDS and hepatitis B), hospitals instituted various forms of Universal Precautions (UP) for blood and Body Substance Isolation (BSI) for other body substances. The concept was formalized by the OSHA Blood-borne Disease Standard in 1992. This was radically different from the historical isolation concept—it was procedure driven rather than diagnosis driven. The concern addressed was that the HIV status of patients was seldom known regardless of diagnosis. Thus patient care practices had to be predicated on the possibility that blood and body substances might be infectious and precautions must be based on that potential exposure. Barriers were utilized to prevent mucous membrane contact with all such blood and body substances with all patients being considered potentially infectious. In the late 1990s, the concepts of UP and BSI were combined under the heading of Standard Precautions. This concept primarily protects the healthcare worker from patients as a containment measure. However, it is also considered to have some value in protecting patients from potentially infected healthcare workers because, as we shall see, the protective barriers can work in both directions.

### Transmission Based Precautions

The new Standard Precautions now dominate infection control practice, but the need for other diagnosis-related practices continues. These categories are now called Transmission Based Precautions and include:

#### *Airborne precautions*

Airborne precautions are for respiratory diseases such as tuberculosis and legionellosis, which are actually airborne (particles $<5\mu m$ and survive in air) and require special room ventilation.

#### *Droplet precautions*

Droplet precautions are for diseases which require mucous membrane contact but are not technically airborne (particles $>5\mu m$ or do not survive in air) and do not need special ventilation precautions. Examples include mumps, chicken pox, and influenza.

#### *Contact precautions*

Contact precautions are for diseases (particularly enteric diseases) spread by hand contact and requiring special handwash precautions.

## CONTAMINATION CONTROL PRACTICES

### Containment

Environmental interventions for nosocomial infection prevention remain controversial. However, there are standard practices associated with both containment and exclusion that are ingrained into the healthcare facility and need no further justification. For containment the following practices are recommended and are designed to protect the worker from the patient:

*Handwashing*

The emphasis is on washing hands *after* each patient contact. Hands should also be washed after visits to the restroom or after any activity resulting in visible soiling. Compliance with this basic concept remains low and ways are sought to make the process more convenient and less burdensome. For example, waterless alcohol-based solutions provide rapid reduction of microbial flora on hands. Such products are widely used in Europe but have not caught on with healthcare facilities in the United States.

*Waste materials*

Linens and disposable waste or reusable items should be bagged for containment before being removed from the room for disposal or reprocessing.

*Personal protective garb*

Personal protective garb (gloves, gowns, masks, respirators) is worn to protect the worker from the patient. Surgical masks have traditionally been worn to protect patients from nasopharyngeal expulsions of the worker. They are, therefore, more appropriate for exclusion than for restraint. The OSHA tuberculosis standard recommends NIOSH-approved respirators for personnel protection.

*Room ventilation*

For containment purposes, the room should be under negative air pressure and exhausted to the outside to prevent escape of aerosols to surrounding rooms or corridors.

*Decontamination of soiled items*

All items, whether disposable or intended to be reprocessed, should be treated physically or chemically to render them free from pathogens before handling by other personnel.

## Exclusion

The same concepts used for containment can provide barriers for exclusion when the patient needs to be protected rather than contained.

### *Handwashing*

The emphasis is on hands being washed *before* each patient contact although good handwashing practice described under containment also should be carried out.

### *Patient items*

Items to be used by the patient should be clean or sterile. There is less emphasis on containment of used items.

### *Personal protective garb and equipment*

The same items (gowns, gloves, masks) used to protect the wearer are now used primarily to protect the patient. Surgical masks can be protective in this instance by containing the expulsions of the workers wearing them. NIOSH-approved respirators should be worn to protect the worker.

### *Room ventilation*

The room must now be under positive pressure (practiced routinely in surgical suites) to prevent inflow of air under doors from corridors or other areas of the building. Filtration of incoming air provides extra protection for immune-compromised patients.

### *Visitor policy*

Visitors should be instructed to utilize the same patient protection practices as healthcare workers. The psychological benefits of visitors generally outweigh the infection risks as long as sensible precautions are exercised.

## REFERENCES

A very comprehensive book on the subject of nosocomial infections is:

Mayhall, G. C., (Ed.) 1996. *Hospital Epidemiology and Infection Control.* Williams and Wilkins. Baltimore, MD.

Other important guidelines related to intervention practices include:

OSHA, 1991. Blood-borne Disease Standard. *Federal Register* Vol. 56, No. 235, Friday, Dec. 6, 1991, pp. 64175-64182.

CDC, 1994. Guidelines for preventing the transmission of *M. tuberculosis* in health-care facilities. *MMWR* Vol. 43, No. RR-13, Oct. 28, 1994.

AJIC, 1996. "CDC Guidelines for Isolation Precautions in Hospitals." *Am. J. Inf. Control,* 24: 24-52.

# Section IV

# PRODUCTS USED AND CONSUMED BY PEOPLE

# Chapter 11

# CONSUMER PRODUCT SAFETY

The vast array of products available to the American consumer presents a particularly perplexing challenge to the concept of human health and the environment. In a free market economy, there is a constant bombardment of hucksterism designed to convince the public to buy, use, and consume this never-ending parade of products. Because liability for harmful defects lies with the seller, people assume that such products, when used as intended, will do the consumer no harm. However, as much as we would like to believe that our lives in this consumer-oriented society can be risk free, we know that isn't the case. In fact, all products carry some combination of risk and benefit, and our use or consumption of them becomes a very complex tradeoff.

For environmental health professionals it comes back to our previously discussed "net societal benefit." Thus, risk assessment, risk/benefit analysis (or risk/tradeoff analysis), and risk communication are all part of the formula.

The concept of combining engineering, legislative, and educational approaches is very pertinent to consumer product safety. It is clear that both public policy and individual decision making come into play. It is also clear that the "outrage factor" demands public policy decisions to protect people from hazards they can't control. Meanwhile, individuals will continue to put themselves at risk due to human foibles and human nature in making decisions about their own lifestyles. Thus, tobacco use, poor dietary habits, and lack of exercise will continue to take their toll on human health, while the same public demands policy actions to protect them from any perceived hazards of the products they use or consume.

## PRODUCT SAFETY CONCEPTS

A closer look at some of the concepts reveals the difficulty of coming up with any magic solution. Scientific uncertainty and the pervasive effect of divergent value judgments leave us with no definitive solutions to defining or controlling risks related to product safety for the population at large. The answer that "it all depends" may be unsatisfying to public health professionals but perhaps best sums up the reality.

**Risk Assessment**

Many books and many models have been devoted to the goal of assessing risk for a given product. The bottom line may be that an overall probability (e.g., lifetime cancer risk from drinking chlorinated water) can be calculated. But how realistic or pertinent is that probability to any given individual? Obviously the risk is much higher for an individual who drinks a lot of chlorinated water and has a genetic disposition to cancer, and much less for an individual who doesn't drink chlorinated water and has no such genetic disposition. The same applies to any other exposure to a product. The major public policy decision involves the risk which can be tolerated by society as a whole. The individual policy decision involves the risk that individual is willing to accept.

**Risk/Benefit Analysis (Risk/Tradeoff Analysis)**

The related question has to do with the tradeoff between perceived benefit and the assessed risk. It is a value judgment which again is made on both the societal and individual levels. Be it the perceived pleasure of alcohol consumption or the convenience of a personal automobile, individuals will often accept a known high risk as justified by that perceived benefit.

**Risk Communication**

I have previously mentioned the failure of environmental health professionals to communicate effectively with the public even when scientific evidence is relatively certain. Try as we must to improve this performance, the individual variances and perceived benefits will always hinder this effort. At least part of the problem of the communication difficulty stems from poor public understanding of probability. Thus, people will protest chlorination of drinking water because they fear the one-in-a-million probability of excess cancer from trihalomethanes, while continuing to spend money on lottery tickets, which promise an even lower probability of hitting a jackpot.

**Net Societal Benefit**

Defining an optimum standard, maximizing benefits while minimizing risks for the greatest number of people, may be elusive and difficult to achieve but should remain the principal objective of environmental health professionals into the twenty-first century.

## ENGINEERING, LEGISLATIVE, AND EDUCATIONAL APPROACHES

### Engineering Approaches

Designing products to be failsafe to the extent possible is certainly a worthwhile objective. Whether dictated by legislative fiat, driven by competitive marketplace realities, or motivated by liability issues, technical applications that reduce hazard, yet require little individual decision-making are vital to consumer safety. From childproof caps on cleaning products to shockproof electric appliances, such engineering has been developed for all of the above.

### Legislative Approaches

Federal- or state-mandated standards for product design or use are also necessary for products that cannot be made totally safe. Examples are particularly pertinent for tobacco products and automobiles.

*Tobacco sales to children*

Tobacco products are inherently hazardous when used as intended. Clearly, political considerations have made a total prohibition of these products unattainable, yielding to the argument that personal choice should prevail. Mandated "Surgeon General" warnings have been used as a shield by the tobacco industry but have not been effective in preventing underage children from becoming addicted before they are capable of exercising that freedom of choice. Laws to prohibit underage sales are difficult to enforce, and imposition of high taxes to reduce consumption is a legislative approach which continues to be vigorously opposed by the tobacco industry.

*Automobile restraints and speed limits*

The personal automobile is deeply engrained in American culture with the benefits of convenience and freedom valued very highly. Great strides have been made in making cars safer, from stability to tire design to safety glass. Highways are also designed for greater safety. Fatalities per million miles driven are significantly lower on the divided highways of the interstate system than they are on two-lane roads.

Mandatory seatbelt use has been enacted in many states with varying degrees of enforcement and effectiveness. Overall seatbelt use has risen from about 15 percent to greater than 50 percent in most locations where such laws are enforced. This rise is credited with significant injury and fatality reduction. The airbag installations mandated in the 1990s are a passive safety feature, unlike seatbelts, thus circumventing the compliance issue. However, airbags have generated some controversy due to injuries to young children. Specially designed, rear-facing child seats are recommended

When national speed limits were reduced to 55 mph during the energy crunch of the 1970s, that move was credited with reducing highway fatalities significantly. Other factors undoubtedly contributed to the reduction, including safer cars and roads and reduction of alcohol use by drivers. In the 1990s, the 55 mph limits have been discarded and speed limits up to 75 mph now prevail on freeways in most states. Highway fatalities are also rising, but the effect of speed limit increases remains unclear due to more miles driven and continued improvements in the other factors. Again, establishing a speed limit based on an "acceptable" number of fatalities becomes a value judgment pitting the lives saved by lower speeds against the perceived convenience and saved time of higher speeds.

### Educational Approaches

There is no doubt that advertising sells products, even those which may be inherently hazardous such as tobacco. Educating people about the hazards of products they encounter remains a difficult public health task, with government advertising budgets minuscule compared to those of industry. Nonetheless, those efforts must be enhanced to play a vital role in protecting health. More effective risk communication remains one of the major challenges for environmental health professionals for the twenty-first century.

## DANGER OF SUBSTITUTING FOR HAZARDOUS PRODUCTS

The trend toward prevention of pollution or hazards rather than end-of-pipe solutions is an important development in environmental health. However, it should be recognized that there may also be costs associated with not using or substituting. Just as few would now advocate return to the horse-and-buggy era to avoid the pollution and injuries inflicted by automobiles, we must exercise caution relative to other products.

For example, asbestos had important benefits as a fire retardant insulating material. Fiberglass, which is now widely used as a substitute, may prove to be at least as hazardous as asbestos in the long run. Similarly, reducing chlorine use due to fears of trihalomethanes could result in increased infectious disease rates beyond any hazard of the trihalomathanes. The lesson to be learned is that there may be a price to pay for not using a product, and a seemingly safer alternative may prove to be a hazard in its own right. Things are never as simple as they seem.

## CATEGORIES OF HAZARD

There may be some degree of risk, however small, associated with almost any consumer product. These risks fall into several identifiable categories: A product may be inherently hazardous when used as intended; it may be accidentally hazardous when misused or malfunctoning; or it may be used intentionally with unrecognized or controversial hazards. Some examples are as follows:

### Inherently Hazardous Products

The most obvious example in this category is tobacco products. Indeed, this may represent the only such product that is legally produced, advertised, and sold to consumers in the United States today. Despite its recognized addictive quality, and an annual death toll from its effects exceeding 400,000, the industry and its congressional allies continue to stymie effective controls, even for marketing to children. The 1998 national settlement endorsed by many states has imposed large financial penalties on the tobacco companies to reimburse the states for healthcare costs resulting from tobacco marketing. Emphasis is placed on restricting sales to minors. The FDA retains regulatory control over nicotine. Separate settlements with several other states, including Mississippi and Minnesota, also impose large financial penalties and restrict advertising (such as billboard location and movie endorsements). However, many public health advocates believe that settlement conditions are not sufficient to discourage teenage or adult smoking. They fear that costs will simply be passed on to smokers in the United States and abroad. The huge death toll from tobacco products continues to be an embarrassment to public health professionals as they seek to regulate other products which exact only a fraction of the tobacco-related toll.

Some people argue that alcohol products should also be in the inherently dangerous category. There is no doubt that abuse of alcohol is responsible for multiple health and social problems. However, alcoholism is a specific disease afflicting susceptible individuals. There is evidence that moderate use (no more than two drinks per day) may actually reduce the risk of heart disease and stroke. High taxation, restricted advertising, and prohibition of sales to minors are all appropriate legislative controls for this product, and educational efforts such as those espousing designated drivers may be responsible for the social change that is reducing the number of alcohol-related traffic deaths.

Firearms are another product which could arguably be considered inherently hazardous. Guns are meant to shoot bullets; they contribute significantly to the growing problem of violence in America. Between 1968 and 1991, firearm fatalities rose about 60 percent (from 24,000 to 38,000) and were on a pace to exceed traffic fatalities by the year 2003 until the recent reversal of traffic death declines (possibly speed limit induced). As with tobacco, a powerful gun lobby has impeded efforts for meaningful gun control, and firearms are frighteningly numerous and available even to minors and mentally impaired individuals. Several cities have initiated legal action against the firearms industry similar to suits against tobacco companies, under the premise that those companies should pay for at least some of the carnage inflicted by their products. The fate of those suits remains to be determined as this is written.

A few natural food products are inherently poisonous (not counting high-fat, high-salt or high-sugar products, which have their own controversial roles in American dietary habits). The primary example is poisonous varieties of mushrooms, which will be discussed in Chapter 13. The deadly mushrooms obviously are not available in retail outlets and need no special legislative controls.

**Accidentally Hazardous Products**

This is by far the largest category of hazardous products, as almost any product could conceivably be misused or could malfunction to create a hazard. Drain cleaners are clearly hazardous if ingested. Warning labels and childproof caps can help prevent such misuse, but containers cannot be made foolproof. In the malfunction category, gas leaks and tire blowouts can cause serious accidents. Engineering design has improved considerably over the years, and the toll taken by malfunctioning products has been reduced steadily. Many federal and state regulations require a redundancy of safety features and label warnings in the array of consumer products. Consumer confidence is also a powerful motivating force, and product liability laws provide an incentive for manufacturers to make products as safe as possible. Congressional efforts to limit the liability of manufacturers pose a threat to these built-in motivations just as limited liability for tobacco companies is a major stumbling block in the ongoing tobacco wars.

**Products with Unknown and Controversial Effects**

Another category of product problems is defined by controversy: Whether or not a hazard actually exists is uncertain. Electromagnetic fields (EMFs) associated with both power lines and appliances such as electric blankets are a good example of this category. A number of highly publicized studies have claimed to show a relationship between some cancers (such as childhood leukemia) and proximity to high voltage power lines. Most epidemiologists, however, dismiss these studies as flawed. Biological feasibility for a cancer effect from EMFs is very tenuous, and numerous studies are also available showing no such effect. The opposite scenario unfolds for the concern with mercury in dental amalgams. Such amalgams were used routinely in dentistry on millions of patients for many years. Whereas there is some biological feasibility for a health effect from release of organic mercury from amalgams, no epidemiological evidence for such an effect has emerged.

Another controversial product hazard is personal scent products such as perfumes, deodorants, aftershave lotions, and hairsprays. Certain individuals have a condition known as Multiple Chemical Sensitivity (MCS)—they react severely to exposure to any of those products. From a scientific perspective, the phenomenon cannot be explained by known biological mechanisms: There is no definitive etiology or latency period for MCS; no definitive diagnostic criteria; and no specific treatment. The major issue raised, therefore, is the appropriateness of any regulatory control. Limitations on personal use of cosmetics would be very difficult to justify or enforce, but some voluntary restrictions have been imposed in workplaces at the request of sensitive individuals.

### Food Additives

The proliferation of chemical additives in the food supply has raised many questions about potential acute and chronic health problems. Those issues and the regulatory approach to addressing them will be discussed in Chapter 13.

### Product Interactions

An obvious product interaction is that of alcohol and automobiles. Fully one-half of traffic fatalities for many years were alcohol related, but a downward trend in that percentage has been observed in the late 1990s (to less than 40 percent by 1998). Societal attitudes toward drinking and driving have shifted for a variety of reasons. A vigorous publicity campaign launched by Mothers Against Drunk Driving (MADD) and strengthening of penalties for conviction are credited for the change. Increasing total traffic fatalities in the late 1990s are attributed partially to higher speed limits, but they may also be due to another interaction; that of an increasing number of heavy trucks and sport utility vehicles whose high bumpers cause extensive damage in collisions with smaller vehicles.

## CARCINOGENICITY AND CONSUMER PRODUCTS

One of the difficult safety issues to assess for any product used or consumed by the public is its potential for causing cancer. The long latency period and numerous confounding factors make epidemiological evidence difficult to evaluate. Animal testing usually provides the best indication of potential carcinogenicity for any substance, but extrapolating from one species to another—humans—is uncertain, particularly when even different rodent species or different genders within a species yield different results. Of necessity, high concentrations must be fed (or aerosolized or injected) to reduce the number of test animals needed and to adjust for their shorter lifespans. This in turn raises questions about applicability to humans. Animal testing is expensive and increasingly opposed by animal rights activists.

Mutagenicity screening tests are often used to reduce the number of substances needing animal testing. The Ames Test (developed by Dr. Bruce Ames) is often used for such screening. Chemicals are tested for their ability to reverse mutations in cultures of *Salmonella* bacteria. Because there is a high correlation between mutagenicity and carcinogenicity, only chemicals demonstrated to be mutagenic are further tested in animals. The fear that everything can be shown to have carcinogenic properties if tested in high enough concentrations against enough rodent species is unfounded. In practice, about 90 percent of chemicals tested are not carcinogenic, and only those with suspected carcinogenic properties are even evaluated.

Further difficulty arises in setting standards based on extrapolating animal testing results to humans. Standards are often set at a level expected to cause no more than one excess case of cancer per million exposed persons. (Approximately one of every three people will eventually die of cancer regardless of these standards. This percentage rises as life expectancy rises due to reductions in heart disease mortality and increased control of infectious diseases.) Once more we are faced with a "net societal benefit" question as we cannot always predict the consequences of not using the suspected carcinogen nor are we sure of the long-term effects of a substituted substance.

Various schemas have been proposed for comparing the relative hazards of substances shown to be carcinogenic. Dr. Ames has proposed using an index called the Human Exposure/Rodent Potency (HERP) Index in which the daily dose (mg/kg) which produces tumors in 50 percent of test animals during their lifetime is compared to the daily human dose (also in mg/kg). Table 11-1 summarizes some of the results of that comparison. Dr. Ames suggests that some of the publicized cancer risks (e.g., chloroform in chlorinated drinking water) are much less of a hazard than some common daily intakes such as beer, wine, or particularly, sleeping pills.

Table 11-1

HERP Index Examples

| Daily Human Exposure | HERP %* |
|---|---|
| Tap water (1 Liter) | 0.001 |
| Bacon (100 grams, fried) | 0.003 |
| Peanut butter (32 grams) | 0.03 |
| Beer (12 oz.) | 2.8 |
| Wine (250 mL) | 4.7 |
| DDT (Dietary intake) | 0.0003 |
| Diet cola (12 oz. saccharin) | 0.06 |
| Phenobarbitol (sleeping pill) | 16.0 |

*Lifetime daily human exposure (in mg/Kg) as a % of dose (also in mg/Kg) required to produce tumors in 50% of test rodents. (Adapted from Ames, et al. *Science* 236:271-279, 1987)

## REGULATION OF CONSUMER PRODUCT HAZARDS

At the federal level a variety of agencies are charged with responsibility for consumer protection. Most of these have been mentioned previously, but the following summary may help to sort out the sometimes confusing jurisdictional picture:

### Food and Drug Administration (FDA)

This is the agency which writes rules for and enforces legislation relative to substances and devices intended for human consumption or contact. Thus, the food supply, drugs, cosmetics and medical devices fall under the FDA. The battle over the FDA's right to regulate nicotine in tobacco products as a drug points out the political implications of basic consumer protection. The FDA often has been criticized for being too slow and bureaucratic in its approval process, but supporters argue that thorough review is necessary for meaningful decision making, often citing the Thalidomide experience of the 1960s as an example. The FDA withheld approval of Thalidomide, a drug approved in Europe that was subsequently shown to cause major birth defects in the offspring of women who had been exposed. That action was credited with preventing thousands of American babies from being born with deformities.

### Environmental Protection Agency (EPA)

This agency was created under the National Environmental Policy Act (NEPA) of 1970 as part of the environmental protection movement. It has jurisdiction for writing and enforcing rules for the pollution control legislation passed by Congress. Air pollution, water pollution, toxic substances, pesticides, and hazardous waste are all regulated by the EPA. Occasionally, the lines between EPA and FDA are blurred, as in the use of pesticides on food crops or the use of germicides on medical devices. For example, the EPA sets the tolerance limits for pesticide residuals on food products, but the FDA monitors the commodities and has authority to confiscate products or order their withdrawal from the marketplace.

### Centers for Disease Control and Prevention (CDC)

This agency, based in Atlanta, Georgia, is responsible for the surveillance of disease and investigation of outbreaks. It prepares authoritative guidelines in many areas but does not have enforcement authority.

### National Institutes of Health (NIH)

Based in Bethesda, Maryland, the NIH is the research arm of the nation's health establishment. It conducts internal research and funds external research through a peer review system. It includes a variety of specialized institutes relating to heart disease, cancer, infectious diseases, and many others. It does not have regulatory or enforcement functions.

### The Consumer Product Safety Commission (CPSC)

The CPSC was created specifically to deal with consumer issues which may fall between the cracks of other agencies. For example, it has been responsible over the years for banning freon-based aerosol cans (for ozone layer protection), assuring that only fire retardant materials are used in children's sleepwear, and banning formaldehyde-based urea foam insulation materials. The agency has been criticized for being politicized and oriented primarily to the interests of the prevailing administration in Washington.

### Other Agencies

Additional federal agencies, such as the Department of Agriculture (USDA) and the Department of Energy (DOE), have some regulatory authority for public health issues. However, because they are advocates for the industries they regulate (agribusiness and electric power), they have less credibility as consumer protection agencies.

In addition to the federal agencies described above all states have comparable state agencies, some of which may have more restrictive laws on consumer issues.

## CONSUMER PROTECTION REGULATIONS

Some of the important federal legislation which has influenced consumer protection in the latter half of the twentieth century is summarized below:

### The Delaney Clause

Passed in 1958 as part of food safety legislation under the FDA, the Delaney Clause was one of the most widely publicized consumer protection laws until it was replaced by the "negligible risk standard" in the Food Protection Act of 1996. The Delaney Clause stated that "no substance demonstrated to be carcinogenic by reasonable tests on man or animals may be present in or added to the food supply." It applied only to deliberate additives, not contaminants or natural ingredients. Even pesticides were covered by other legislation. It applied only to cancer, not other hazards. It was widely criticized because its zero tolerance provision—"no substance"—proved to be impractical and unenforceable as technology improved for the detection of chemical traces at the parts per billion or parts per trillion level. It was also inconsistently applied. In fact, it was actually invoked to ban only two chemicals in its 38 years. Those two, diethylstilbestrol (DES) and cyclamates, will be discussed in Chapter 13. The negligible risk standard differs from the Delaney Clause in that it recognized that minute quantities of suspected carcinogens may be present without sufficient hazard to warrant a ban.

### The Generally Recognized as Safe (GRAS) List

Also passed in 1958 under the FDA, the GRAS list applies to chemicals in food that are not deemed to be carcinogenic. It requires that any new chemicals proposed as food additives be tested for safety (toxicity and teratogenicity) before approval. Chemicals previously added to food with no indication of human health effects do not need to be tested—are grandfathered in—unless subsequent evidence indicates harm to humans. Unlike the Delaney Clause, the GRAS list does not have a zero tolerance provision. The hazardous level is determined by animal testing, and a safety factor (often 100 times higher) is added to establish the legal limit, or No Observable Adverse Effects Level (NOAEL).

### Toxic Substance Control Act (TSCA)

This Act was passed in 1976 and assigned to the EPA. Unlike the Delaney Clause and GRAS list, which applied to the food supply, the TSCA applies to all chemicals (other than in food) which are newly introduced or proposed for new uses. It requires a premanufacturing notice (PMN) for such intent, 90 days before marketing. The EPA then has the responsibility for determining whether the chemical might pose a health hazard to the public. If so, safety testing is required. Enforcement of TSCA has been hampered by budgetary restrictions and political influence making its overall effectiveness somewhat questionable.

### The Federal Insecticide, Fungicide and Rodenticide Act (FIFRA)

A pesticide regulation at the national level was first passed in 1947, but the major provisions of FIFRA were inserted into major 1972 amendments. Like the GRAS list and TSCA, FIFRA is a "front end" Act, requiring testing before the pesticide is marketed. Pesticides will be more extensively examined in Chapter 15. They differ from other chemicals in that they are intentionally toxic to their target species and often may be inadvertently toxic to humans and other species as well. The testing is designed to determine how a pesticide can be applied safely. Each one is usually approved only for specific applications at specific times (for instance, sprayed on fruit trees only before the blossoms appear). The labeling requirements are very specific in terms of approved uses and first aid instructions for accidental exposure. FIFRA is enforced by the EPA, although some pesticides (such as disinfectants for medical devices) are also regulated in part by the FDA.

**REFERENCES**

Useful references in the area of consumer product safety and pertinent regulations include:

American Association for the Advancement of Science. 1987. "Risk Assessment Issues." (6 articles). *Science.* 236: 267-302.

Dyer, K. S. and K. Sexton. 1996. "What can research contribute to regulatory decisions about the health risk of Multiple Chemical Sensitivity." *Regulatory Toxicology and Pharmacology* 24: S139-S151.

Findley, R. W. and Farber, D. A. 1992. *Environmental Law in a Nutshell.* West Publishing Co. St. Paul, MN.

Graham, J. O. and J. B. Winer (Eds.) 1995. *Risk vs. Risk Tradeoffs in Protecting Health and the Environment.* Harvard University Press. Cambridge, MA.

# Chapter 12

# FOOD SAFETY: BIOLOGICAL AGENTS

In the next two chapters I will explore the safety issues surrounding the one product which all Americans consume every day, namely food. In this chapter I will discuss biological (infectious) agents, and in Chapter 13 I will discuss problems associated with chemicals in food. Infectious agents spread through food are acquired via the ingestion route and not surprisingly include many of the same agents implicated in waterborne disease. However, there are several differences which make food a more dangerous vehicle of transmission than water. Whereas water is provided to communities from a central source with a controlled treatment facility and distributed through closed mains to the consumer, food comes from multiple sources and goes through multiple handling steps including those of final home or restaurant preparation. Thus, there are multiple opportunities for contamination and mishandling. In addition, although water is a carrier of pathogens, it does not provide a good medium for growth and multiplication. Many foods, on the other hand, particularly those which are proteinaceous, lend themselves to proliferation to high concentrations, especially when temperatures are in the mesophilic (near body temperature) range.

Relative to other countries, the food supply in the United States has been considered safe. However, in recent years a number of well-publicized outbreaks have increased awareness that this safety presumption cannot be taken for granted. Changes in sources, production methods, distribution, and consumption patterns all have implications for food safety, prompting passage of legislation in 1996 strengthening food safety regulations. Some of the trends influencing food safety are as follows.

Trade agreements (such as NAFTA), and increased global marketing and distribution patterns, have resulted in a higher percentage of imported foods in the United States. During the winter months, more than 50 percent of fresh produce is imported, mostly from Mexico and other Latin American countries. Although the trade agreements include provisions aimed at equalizing sanitary harvesting and processing standards, those provisions have been difficult to enforce and sanitary quality has been called into question. Several disease outbreaks have been attributed to that problem although it is a difficult link to prove definitively. In the 1990s, a diarrheal outbreak attributed to an emerging protozoan agent named *Cyclospora cayetanensis* (also an example of a newly emerging pathogen) has been associated with Guatemalan raspberries, and a hepatitis A outbreak has been associated with Mexican strawberries.

Even in the United States, production changes may be influencing food safety. For example, cattle are raised in large, crowded feedlots with animals from different sources coming together, increasing the probability of spread of pathogens. Ground beef is processed in central facilities, creating the potential for a pathogen such as *Escherichia coli* 0157:H7 to be mixed throughout a large quantity of ground beef, which is then distributed over a multistate area. In 1997 there was a much publicized recall of more than 25 million pounds of ground beef after discovery of that pathogen in a batch of product.

There is also a continuing trend for Americans to consume more meals away from home. Breakdowns in sanitation occurring in restaurants affect more people and are more likely to come to the attention of authorities than those occurring at home. They are also subject to greater supervision.

## SCOPE OF FOOD-BORNE DISEASE

Typical acute food-borne illness is something which almost all Americans experience during their lifetime, usually more than once. The total number of such illnesses is problematic, because the great majority are never diagnosed. Illness is often mild and transient with full recovery within a day or two. Thus the term "24-hour stomach flu," "food poisoning," or something similar is often used to characterize the illness. Typical symptoms include some combination of diarrhea, nausea, vomiting, fever, and, occasionally, headache or other symptoms. Acutely miserable as the victim might be, the worst is often over before medical attention can be sought. It was estimated by the U.S. Department of Agriculture (USDA) in 1998 that somewhere between 6 million and 30 million such illnesses occur annually in the United States, although the number of confirmed isolations of enteric pathogens is only a fraction of that total. Many food-borne pathogens are nationally reportable but the infrequent confirmations and inconsistencies between states in reporting diligence make it difficult to interpret those numbers. The overall fatality rate for food-borne illness is estimated to be less than 0.1%. However, with so many cases occurring, that low rate translates to about 10,000 fatalities every year in the U.S.

## TYPES OF FOOD-BORNE ILLNESS

In subsequent sections I will discuss several categories of food-borne illness caused by infectious agents. They include the true infections, in which the agent establishes itself in the host and multiplies there with illness resulting from irritation of mucosa and/or production of toxins by those organisms. True infections may be caused by bacteria which often can be transmitted from person to person (secondary transmission) as well as by food and water. Viruses, protozoa, and helminths are other types of organisms that are responsible for true food-borne infections. I will discuss examples of all of those.

The second major category is bacterial intoxications, in which an exotoxin is metabolized by the organism while it is growing and multiplying in the food it has contaminated. It is ingestion of the food containing the toxin which causes the illness. The organism itself may not even be present. For example, in staphylococcal food-borne intoxications, where the organism may be killed by cooking, the heat-stable toxin survives and causes illness without any trace of the pathogen in the victim or in the food.

A third form of food-borne microbial illness is toxicoinfections. They are illnesses caused by bacteria which are noninvasive (they do not establish themselves and multiply in the host) but which produce toxins while passing through the intestines. Examples include *Clostridium perfringens* and *Clostridium botulinum* (in infants).

## Bacterial Food-borne Infections

### *Salmonellosis*

The most frequently confirmed of all food-borne diseases are those caused by various species of salmonella. More than 2,000 serotypes of this genus have been identified (about 200 in the U.S.) and new developments in DNA fingerprinting allow sub typing, which can confirm common source etiology for food-borne outbreaks. More than 45,000 confirmed isolates of salmonella are reported in a typical year, although this is suspected to be only the tip of the iceberg. Several large outbreaks of salmonellosis, which typify how this agent can be transmitted, have been investigated in recent years. One of those occurred in the Chicago area in 1985. It was associated epidemiologically with fluid milk from Jewel Dairy, a large supplier of dairy products in the Chicago Metropolitan area. More than 15,000 cases were confirmed with more than 150,000 suspected. No specific defects were identified in the modern facilities of this dairy and the chain of transmission can only be speculated. The pasteurization process is used as the primary environmental intervention strategy for the safety of dairy products. This process kills pathogens with a considerable margin of safety and records indicated no glitches in that process. Postpasteurization contamination or a cross-connection to unpasteurized milk were the most likely culprits in that outbreak.

An even larger outbreak, believed to be responsible for more than 250,000 cases nationwide, was reported in 1994. That outbreak was associated with ice cream distributed nationally by Schwan's, a frozen foods company in Marshall, Minnesota, and may be the largest single food-borne outbreak recorded in the United States.

Again, the chain of events is somewhat speculative, but in this outbreak a specific deficiency was identified.

The ice cream plant in Marshall received prepasteurized ice cream mix from several pasteurizing plants in Minnesota. Again, the pasteurization records indicated sufficient time and temperature for certainty that the pathogen (in this case *S. enteriditis*) was killed. However, some of the mix was transported to Marshall in a refrigerated stainless steel tanker truck which had previously hauled unpasteurized raw egg mix. Eggs are a common source of salmonella. The trucks were steam sanitized between uses. However, one theory holds that there were small cracks in the truck liner. When the egg mix was pumped into the truck under pressure the cracks opened and contaminated egg mix seeped into them. The truck interior was not pressurized during the sanitization process so the cracks may have been effectively sealed off from the steam. When the pasteurized ice cream mix was introduced, again under pressure, the cracks reopened exposing the mix to the contaminated egg. It is also possible that some of the valves in the system were not adequately steam sanitized. In addition, it was reported that the truck was left outdoors overnight in warm weather between the emptying of the egg product and the sanitization step, which could have increased the number of organisms significantly.

The epidemiological chain for salmonellosis is typical of many bacterial food-borne infections. The agent has its reservoir in the intestinal tract of humans and many wild and domestic animals including poultry, hogs, cattle, dogs, and turtles. Thus, meat and poultry products are common sources as are infected food handlers. *S. typhi*, the causative agent of typhoid fever, is known for the importance of the human carrier state. Escape is through the intestinal tract and meat and poultry products can be contaminated during the butchering and processing steps. Human food handlers are also a likely source of contamination. Transmission occurs when contaminated foods are left in the mesophilic growth range for sufficient time to reach the infectious dose. It has usually been reported that at least $10^3$ to $10^4$ organisms are needed to initiate infection in a healthy host. Thus proteinaceous foods supporting such multiplication are the most dangerous. However, recent evidence indicates that salmonella can be associated with fresh produce. For example, canteloupes and alfalfa sprouts have been implicated in outbreaks, extending the need for sanitary precautions to a wider range of foods. The population at risk is almost universal and individuals can be infected more than once due to the wide range of serotypes.

**Hazard Analysis Critical Control Points (HACCP)**

The Hazard Analysis Critical Control Points (HACCP) concept has assumed increasing importance in food safety in the 1990s and has been specifically incorporated into the strengthened rules promulgated by the 1996 Food Safety Legislation. For any food processing operation, a step-by-step analysis must be carried out to identify points for possible introduction or amplification of pathogens.

Written procedures are required to assure that pathogens are destroyed or excluded at each of these points. HACCP implies self-inspection, reducing the cost of regulatory oversight.

## Other Bacterial Food-borne Infections

While salmonellosis is the most frequently reported bacterial food-borne infection and is typical of the chain of events leading to disease, several other important pathogens should be noted.

### *Shigella species*

Shigella are second to salmonella in total isolates most years, accounting for about 25,000 isolates per year. Humans are the major reservoir. The organism does not colonize food animals, which probably accounts for the smaller number of cases relative to salmonella. However, it generally requires a much smaller infectious dose (as few as 10 to 100 cells).

### *Vibrio species*

This group includes the cholera organism (*V. cholerae*), which was discussed in Chapter 3. It also includes *V. parahaemolyticus* which has its reservoir in ocean sediments and is transmitted through undercooked fish, and *V. vulnificus*, which is associated with eating raw shellfish. *V. vulnificus* causes extremely dangerous illness and fatalities in persons with preexisting liver disease.

### *Escherichia coli*

Enterotoxigenic *E. coli* have been recognized for many years as a cause of "traveler's diarrhea" in many parts of the world. However, the emergence of enterohemorrhagic strains, especially the serotype 0157:H7 in the 1980s, has been particularly important in emerging food-borne disease problems. These strains produce a shiga toxin (possibly transferred naturally from a *shigella* species) which can cause an often fatal complication called hemolytic uremic syndrome (HUS) in children under the age of five. A highly publicized outbreak in 1993 in the Pacific Northwest was a major factor in focusing the need for more stringent regulations. That outbreak was associated with the Jack in the Box fast food chain. It was traced to undercooked hamburger which had originated from feedlot cattle and was processed in a single facility and distributed widely to retail outlets in a number of states. The outbreak caused more than 500 cases, and there were four fatalities from HUS in young children. *E. coli* 0157:H7 was made reportable nationally through CDC in 1994 and more than 2,000 isolates have been confirmed each year since then. The common occurrence of *E. coli* 0157:H7 in hamburger spurred the approval of beef irradiation in 1997. Another innovative suggestion is that infection of cattle might be reduced if the animals were fed hay instead of grain in the weeks just prior to slaughter. In addition to hamburger, *E. coli* 0157:H7 has been associated with a variety of other foods, including unpasteurized apple juice in California and radish sprouts in an outbreak in Japan in 1996 which caused more than 9,000 cases and killed seven children.

*Listeria monocytogenes*

Listeriosis is noteworthy in that the organism can grow at very low temperatures and also is relatively resistant to heat. It has an environmental reservoir (mud, silage, water). Associated with soft cheese products, it is particularly threatening to immunosuppressed individuals and pregnant women. A nationwide outbreak of listeriosis was reported in the United States in late 1998. The outbreak was traced to processed hot dogs and resulted in a substantial product recall.

*Campylobacter jejuni*

Campylobacteriosis is a very common infectious disease believed to be responsible for about 10 percent of cases of diarrheal illness worldwide. It has its reservoir in numerous animals, notably poultry, cattle, and domestic pets. Although it has not received as much publicity as salmonellosis, it may cause as many or more cases of food-borne illness in some states in the United States.

## INTERVENTION MEASURES

Intervention steps for salmonellosis are also typical of those for other food-borne bacterial infections. Control over raw products and personal hygiene among food handlers (particularly handwashing) are important intervention steps. Refrigeration to keep foods out of the growth temperature range is also vital. Terminal cooking before ingestion will kill salmonella and other bacterial, viral, and parasitic agents. It will not, however, destroy heat-resistant toxins such as the staphylococcal toxin. General sanitation in food preparation areas, including sanitization of food contact surfaces and vermin control, is also important. Irradiation, which was discussed in Chapter 7, is thought by some to be a necessary future development to assure pathogen free status for many foods.

### Food-borne Viral Infections

Although food-borne viral infections are true infections (the agent establishes itself and multiplies in the host), they differ from bacterial infections in several important ways. Viruses multiply only in living cells and therefore do not replicate in food. Thus, the original contaminating event must provide sufficient virus particles to form an infectious dose. A great variety of foods may be implicated—not only proteinaceous foods—and refrigeration does not prevent the outbreak from occurring. Because viruses are heat sensitive, ordinary cooking prevents disease; outbreaks are associated with foods served cold. Humans are the reservoir, and personal hygiene of food handlers, particularly handwashing practice, is the principal intervention strategy. Virus-associated outbreaks are more difficult to confirm than bacterial outbreaks, although new advances in detection methods such as Polymerase Chain Reaction (PCR) are improving the rate of confirmation. The enteric viruses most often implicated in food-borne transmission are:

*Hepatitis A (also sometimes hepatitis E)*

Numerous food-borne outbreaks have been caused by hepatitis A, an RNA virus in the picornavirus family. Food handlers who are carriers of the virus are frequently the source of infection. Salads and cold sandwiches are often implicated because they require extensive handling in preparation. Shellfish eaten raw have also been associated with outbreaks when harvested from water contaminated with human excreta. Person-to-person transmission is also common.

*Norwalk viruses*

Norwalk viruses are RNA viruses in the calcivirus family. They are notoriously difficult to detect but are believed to be responsible for about one-third of nonbacterial gastroenteritis outbreaks. Humans are the reservoir, and transmission factors are similar to those for hepatitis A. In one typical outbreak, chocolate frosting applied to already-baked cupcakes and donuts (traced to a bakery in Anoka, Minnesota) was responsible for more than 200 cases. The worker who mixed the chocolate frosting manually had been severely ill with diarrhea just prior to his work shift. Personnel policies to prevent ill employees from handling food were an obvious preventive measure lacking in that incident.

## Prions

In 1996 there was a highly publicized outbreak of a new variety Creutzfeldt-Jakob disease (nv CJD), popularly called "mad cow" disease, reported from Great Britain. The agent for this disease is probably a filterable self-replicating protein called a prion. Highly fatal CJD and other degenerative spongiform encephalopathies of humans and other animals were previously thought to be caused by a group of spongiform (slow) viruses characterized by long incubation periods and extreme resistance to heat and other cidal treatments. The scenario for transmission of this nv CJD to humans poses many questions for food safety. The human illnesses (about 21 cases) are postulated to have been transmitted through the ingestion of beef. British cattle were victimized by a major outbreak of Bovine Spongiform Encephelopathy (BSE) starting in 1986. The outbreak was traced to the feeding of a dietary supplement containing traces of neurological tissue from other cattle. Although the practice was discontinued in 1989, cases in cattle continued to be detected into the mid-1990s due to the long incubation period. Exports of British beef were halted and thousands of cattle slaughtered to contain the outbreak. The control measures have apparently been effective in preventing spread of the disease to other countries, and only one human case (in France) has been confirmed outside of Great Britain (as of 1998). The implications of this outbreak are significant in that infectious prions had previously been thought to be confined to a single species (e.g., scrapie in sheep) and not to cross species barriers in a single passage. In the United States, there is a ban on adding any animal tissue to livestock feed.

**Food-borne Helminthic Infections**

There are several types of helminths which can be transmitted to humans through the food supply:

*Trichinosis*

Trichinosis is a disease caused by an intestinal roundworm (*Trichinella spiralis*) whose larvae migrate to and become encysted in muscle tissue. Transmission to humans occurs through the ingestion of animal tissue (meat) containing encysted worms. In the United States this has primarily involved the ingestion of undercooked pork products. The trichina worms are readily destroyed by cooking that raises the temperature of all tissue to 71°C (160°F) or by prolonged freezing (approximately 0°F for 5 days). The cultural preference for well-cooked pork and regulations re-quiring pasteurization of hog feed have reduced the reported incidence in the United States to a very low level (usually fewer than 50 cases annually). Occasional cases are also attributed to ingestion of meat from other carniverous species, such as bears or walruses, or to undercooked sausages or cold cuts not recognized as containing pork.

*Taeniasis*

The tapeworms of cattle and hogs are also implicated in food-borne infection. Humans are the reservoir with cattle or hogs being secondary hosts. Humans then acquire the worms by ingesting undercooked beef or pork contaminated with the eggs or larvae from infected animals. The pork tapeworm (*Taenia solium*) is particularly dangerous, causing neurological complications. The worms can be detected in cattle or hogs by veterinary inspection and, like trichina worms, are readily killed by cooking or prolonged freezing.

### Protozoan Infections

The protozoans previously described in Chapter 3—*Giardia, Cryptosporidium,* and *Amoeba*—are all more commonly spread through drinking water but can also be transmitted in food. In 1996 an emerging protozoan pathogen (*Cyclospora cayetanensis*) was responsible for an outbreak of diarrheal illness affecting more than 1,400 people in the United States. It was apparently transmitted by imported raspberries from Guatemala. The suspected source of the protozoa was an insecticide spray formulated with contaminated water. Raspberries are particularly difficult to clean. Because of the sticky surface of the berries, contaminants cannot be easily removed without destroying the texture of the fruit. All of the protozoa discussed in this section are readily killed by heat. However, they are more resistant to chemical germicides than are bacteria or viruses.

## MICROBIAL INTOXICATIONS

The symptoms of a typical food-borne microbial intoxication are very similar to those of a typical microbial infection. The incubation period is shorter (one to three hours as opposed to 12 to 36 hours for true infections) and there is less likely to be a fever associated with the illness. Because the organisms are often absent from the food when ingested, these intoxications often lack definitive confirmation. The toxins are difficult to identify. The most common of the organisms associated with food-borne intoxication is *Staphylococcus aureus*. Similar symptoms are produced by *Bacillus cereus* toxin. The other major food-borne intoxication is botulism, which is a serious neurological disease which may be fatal if not diagnosed and treated promptly.

### Staphylococcal Food-borne Intoxication

*S. aureus* has a primarily human reservoir and occurs mostly from contamination from a food handler. It is found in the nasopharynx or in boils on the skin rather than in human intestines. It usually requires an opportunity to grow and multiply in the food, and is thus frequently associated with proteinaceous foods. Very importantly, the toxin is heat stable and will survive ordinary cooking temperatures. The primary preventive measures, therefore, are personal hygiene and good food handling practice to avoid contamination of the food, and refrigeration (<4°C) to prevent growth and multiplication.

### *Bacillus cereus*

This intoxication results from a spore-forming bacteria. The spores often survive the cooking process, then germinate and multiply if given a proteinaceous food environment and mesophilic temperatures. The toxins may be either heat stable or heat labile. Outbreaks are often associated with leftover foods which have been cooked and presumed safe. However, if the spores have survived cooking, the organisms will grow and elaborate the toxin. Prevention emphasizes keeping leftovers out of the mesophilic growth temperature range. This organism is more prevalent in Europe than in the United States.

**Botulism**

This disease is caused by an anaerobic spore-forming bacterium (*Clostridium botulinum*) with a soil reservoir. Vegetables are often contaminated during growth and harvesting. The food processing industry relies heavily on the high heat retort process in commercial canning which reduces the probability of a surviving spore to less than one in a billion. However, in home canning, where the controls are not as rigid, spores may survive. The anaerobic environment in the can then allows the spore to germinate and the vegetative organism to multiply and produce toxin. Slightly alkaline vegetables such as beans and mushrooms present the highest risk. Unlike the staphylococcal toxin, however, botulism toxin is heat labile and will be inactivated by ordinary cooking. Thus, food-borne botulism requires an unbroken chain of circumstances, including contamination of the vegetables, inadequate canning temperature, a pH suitable for multiplication of the organism, and ingestion of unheated product, accounting for the small number of cases reported each year. Typically, about 25 cases per year are reported to CDC. In addition to home canned foods, occasional cases are associated with vacuum packed seafood products (usually type E toxin rather than A or B), and even with garlic sauteed in oil.

Another form of botulism, actually causing more cases each year (50-100), is infant botulism. The intestinal tract of the human infant has a higher pH than the adult intestine. Thus, ingested spores (which come from raw products such as honey or corn syrup) may germinate in the anaerobic environment of the intestine and produce the toxin. This is an example of a toxicoinfection which was defined earlier. Prevention of infant botulism is best achieved by assuring that no raw foods, potentially contaminated with botulism spores, be fed to infants.

**Other Toxins Associated with Microbial Action**

In Chapter 13 I will discuss the various mechanisms by which toxic chemicals can enter the food chain. There are several such chemicals which result directly from microbial action. Two such examples are saurine (responsible for scromboid fish poisoning) and aflatoxin.

*Scromboid fish poisoning (saurine poisoning)*

One of the most frequently reported chemical food poisons is saurine. This toxin is associated with scromboid fish, including the frequently eaten staples, tuna and mackerel. The toxin is a form of histidine, which is a decay reaction product of histamine from the muscle tissue of the fish. The role of the microbes is to expedite the decay process. Outbreaks are traced to poor sanitation and lack of refrigeration on the fishing vessels harvesting the catch. Certain bacteria (notably *Proteus* sp.) facilitate the conversion of histamine to the toxic histidine. The toxin is heat stable and even survives the canning process. Thus, prevention must rely on scrupulous sanitation to avoid conamination and on refrigeration to prevent spoilage prior to the canning process.

*Aflatoxins*

Unlike the microbial toxins discussed so far, aflatoxin is more potentially hazardous to humans as a suspected carcinogen than as a toxin. Aflatoxin is metabolized by a fungal agent (*Aspergillus flavus*). It is believed to be an important cause of liver cancer in humans in some parts of the world. It is present in crops stored under humid conditions. Peanuts have been especially susceptible, thus raising concern about aflatoxin residues in products such as peanut butter. The FDA has established a guideline of 20 parts per billion (ppb), although aflatoxin never was covered by the Delaney Clause because it is not a deliberate additive. Prevention is achieved primarily through sanitation and controlled humidity in grain storage facilities.

## REFERENCES

A classic text on the subject of food safety is:

Reiman, H. and F. L. Bryan. 1979. *Foodborne Infection and Intoxication.* (2nd Ed.) Academic Press. New York, NY.

Other useful current references in this area include:

Centers for Disease Control and Prevention. 1996. Foodborne disease outbreaks, five year summary, 1988-1992. *Morbidity and Mortality Weekly Report.* 45(SS5): 1-66.

Council for Agricultural Science and Technology (CAST). 1994. Foodborne Pathogens: Risks and Consequences. CAST- 4420 West Lincoln Way, Ames, Iowa 50014-3447.

Doyle, M. P., L. R. Beuchat, and J. Montville. 1997. *Food Microbiology: Fundamentals and Frontiers.* ASM Press. Herndon, VA.

# Chapter 13

# FOOD SAFETY: CHEMICAL AGENTS

Although the great majority of reported acute food-borne illness is caused by infectious agents, chemicals in the food supply have been the subject of extensive controversy and debate. While acute chemical toxicity is reported annually through the CDC surveillance system and preventive measures instituted, long-term effects are more difficult to evaluate. In this chapter I will review the various mechanisms by which potentially hazardous chemicals get into food and applicable preventive interventions.

## CHEMICAL AGENTS ASSOCIATED WITH NATURAL PROCESSES

In some food products, highly toxic chemicals are present through natural processes. One such example, which has been mentioned previously, is the alkaloid hepatotoxin found in some species of wild mushrooms. Some 90 percent of worldwide fatalities resulting from mushroom poisoning are associated with one species, *Amanita phalloides*. This species is common in Northern California but is also found elsewhere in the United States. The poisonous species cannot easily be distinguished from nonpoisonous varieties by amateur pickers. The best prevention is education to inform the public of the potential hazard and the wisdom of sticking to commercially grown mushrooms.

The chemical most frequently reported by CDC as a cause of food poisoning is ciguatoxin. This toxin is contained in algal dinoflagellates which are part of the food chain. High levels are sometimes found in herbivorous ocean fish, particularly the grouper and snapper varieties. Human ciguatera fish poisoning occurs in areas where fish with high levels of the toxin are consumed. Local concentrations have been detected in Guam, the Virgin Islands, and parts of Florida. The heat-stable toxin is resistant to ordinary cooking and causes both gastrointestinal and neurological symptoms. It can be detected through a stick assay, which is not foolproof because it does not detect all forms of the toxin. A new assay based on cell culture has shown some promise. Saxitoxin is another toxin with similar etiology. It concentrates in shellfish. Like ciguatoxin, it is heat stable and concentrates in the broth of shellfish stews during the boiling process. Reactions to this toxin can be severe and even life threatening.

Another natural chemical which may reach toxic concentrations is selenium, a natural constituent of soil which is taken up by crops, such as wheat. The only prevention is to avoid growing such crops in soil known to have a high selenium content. Selenium is particularly interesting in that it is an essential nutrient in low concentrations and additionally is believed to be an antagonist to arsenic, thus serving as an anticarcinogen.

## ACCIDENTAL CONTAMINANTS

### Pesticide Residues

There are numerous mechanisms through which unintended chemicals may contaminate food. Pesticides are obviously used intentionally, but residues are not intended to remain in or on the food when it reaches the consumer. This issue is particularly controversial because of the uncertain effects of trace residues. The EPA has the responsibility for setting limits on residues remaining on food products following legal use. There have been a number of publicized recalls of products after FDA inspections have discovered residues above regulatory limits. The issue was first publicized in the 1950s when an herbicide called aminotriazole was found in canned cranberry sauce on supermarket shelves shortly before Thanksgiving. The recall—devastating to the cranberry growers, who depend on holiday season sales—was prompted by laboratory tests which showed a carcinogenic potential for the legally used weed killer. In the 1970s another major recall generated publicity when ethylene dibromide (an insecticide used on grain products) was detected in cold cereal. Again, the product was used legally but residues were not intended to remain in the consumer product.

One of the most highly publicized controversies occurred in 1989 over the use of daminozide (Alar), which was used extensively in apple orchards. Alar was a very important chemical for the orchards because it prevented premature ripening, which caused apples to fall to the ground. EPA had considered banning Alar as early as 1980 but had not done so because the agency did not deem the risk to be significant. The publicity erupted in 1989 following a report on the television program "60 Minutes" criticizing the EPA for not taking action. The publicity was extremely damaging to the industry, resulting in a voluntary halt to the domestic distribution of the chemical by the manufacturer. This incident is now looked upon as an example of poor risk communication by EPA in a situation where animal testing was too inconclusive for extrapolation to human cancer potential.

*Illegal Use of Pesticides*

Some incidents of pesticide residues detected in food have followed illegal use of a particular application. In 1985 there were more than 100 cases of acute toxicity reported from consumption of watermelons grown in California. The melons had been treated with a carbamate insecticide (Aldicarb) that was not approved for use on watermelons. In another incident in 1994, Dursban (a pesticide used for roach control) was found in stored grain in a General Mills storage silo in Duluth, Minnesota. The contaminant was traced to illegal use by a contractor trying to save money and resulted in a subsequent criminal prosecution.

**Leaching of Heavy Metals from Cookware Coatings**

Another mechanism for the appearance of acutely toxic chemicals in food is leaching from food contact surfaces, usually under high-acid (low pH) conditions. Examples include the leaching of copper from piping in carbonated beverage machines. Copper piping should obviously not be used for such machines because of the low pH associated with carbonated beverages. Similarly, lead is sometimes found in the glazing of ceramic mugs or other cookware. Acid beverages have the potential for leaching of the lead. One other example is the use of enamel cookware which uses a base metal containing cadmium. When the enamel is chipped, and the base metal exposed, the same potential for leaching by acid foods exists.

**Accidental Substitution**

There have been several incidents reported over the years when the unlikely substitution of a toxic substance, similar in appearance to the intended contents, has caused problems. "The Eleven Blue Men," written by Berton Roueche in 1949, tells a classic tale of such accidental substitution. In this incident, sodium nitrite, which closely resembles sodium chloride, was put into salt shakers in a restaurant in the Bowery section of New York City. (Sodium nitrite is used as a preservative in sausage and cold cuts and was thus available in an improperly labeled container.) The clientele consisted mostly of derelicts who nursed hangovers by heavily lacing hot breakfast cereal with table salt. The result was eleven cases of cyanosis (hence eleven blue men) which had to be treated in the local emergency room. In a subsequent similar incident that occurred in a Chicago delicatessen, there were several fatalities.

Another widely publicized incident of accidental substitution occurred in Michigan in 1973. A company that produced a variety of products somehow managed to package a fire retardant material containing polybrominated biphenyl (PBB) into bags intended for and labeled as a cattle feed supplement, which they also produced. The mislabeled bags were distributed to farmers statewide and the contents unknowingly fed to large numbers of animals. Many of the animals sickened and died or were slaughtered. When the mistake was discovered, there was considerable concern about possible human health effects. Such effects were never definitively proven. Lawsuits by the affected farmers dragged on for many years and cost the company millions of dollars in settlements.

There is no fully dependable mechanism for preventing incidents of accidental substitution. Labeling requirements, fear of bad publicity, and the threat of lawsuits all serve to make people more careful, but the consumer usually depends on blind faith that containers contain what the label indicates and can only be wary of unlabeled or relabeled containers or any suspicious appearance of a product.

**Industrial Discharges**

I have previously discussed toxins in seafood products resulting from natural food chain events (dinoflagellates). Fish can also concentrate toxic chemicals appearing in the food chain through industrial discharge of such chemicals as organic mercury and polychlorinated biphenyls (PCBs).

*Organic mercury*

An incident in Japan in the 1950s alerted the world to the potential problems of organic mercury in fish. Factories were discharging mercury into the waters of Minimata Bay, which also harbored a commercial fishing industry. Mercury was being bioaccumulated in the fish tissue and severe mercury poisoning occurred in many people who consumed the fish. The crippling neurological symptoms were subsequently called Minimata disease. Control over direct discharge of mercury from industrial operations is clearly needed for prevention. However, it is now recognized that traces of mercury can appear in lakes far removed from any such industrial discharge. It is postulated that such contamination may result from airborne transport from remote power plants or municipal incinerators. Stringent emission standards for such sources are needed to minimize this problem. Fish advisories have been issued for many lakes in the United States; these recommend limits on the number of times per month particular species of fish should be consumed.

*Polychlorinated biphenyls (PCBs)*

A similar situation, including fish advisories, exists for PCBs. For many years PCBs were used extensively in many products. They were incorporated into electric transformer oils for their dielectric and fire retardant properties. They were also components of copying paper, ink, paints, and pesticides, among many other products. In 1968 more than 1,800 illnesses in Japan (subsequently called Yusho disease) were reported from consuming rice oil which had been contaminated with PCBs from a faulty heat exchange device. The major symptom was chloracne, a dermatological eruption. A similar incident was reported in Taiwan in 1979.

In 1976 PCB manufacture was banned and a campaign started for removal of the persistent chemical from transformers and other machinery. The FDA set limits on PCBs in fish as early as 1973. Despite the ban, PCBs are still detected in fish from some locations and advisories remain in effect. As with mercury, airborne transport from power plants and incinerators is a suspected source. The human health effects of PCBs appear to be limited to the chloracne symptoms. However, more serious effects are suspected to be related to common comtaminants of PCBs, namely polychlorinated dibenzo-dioxins (PCDDs, commonly called dioxins) and polychlorinated dibenzofurans (PCDFs). Dioxin in particular has aroused heated debate concerning its potential human health effects. It was the suspected cause of health effects in servicemen who had sprayed Agent Orange as an herbicide during the Vietnam War.

It was also the focus of the highly publicized Times Beach incident. Times Beach was a resort town in Missouri with a population of about 2,000 people. In the 1970s the town hired a contractor who utilized dioxin-laced oil to cover unpaved roads. In 1983 the town was basically condemned. Properties were evacuated and bought out by the government because of suspected dioxin contamination. Times Beach became a ghost town, subsequently converted into a state park. This incident is now viewed by many as a gross overreaction to an unproven human health problem, with no evidence that dioxin was responsible for illness in any of the residents of the town.

Actual health effects from dioxin remain controversial. Most scientists now believe that it is a potential carcinogen only at very high exposure levels. It may be a concern for some occupational exposures but is probably insignificant as a dietary contaminant for the general public exposed only to minute traces.

## DELIBERATE CHEMICAL FOOD ADDITIVES

Chemicals that are deliberately added to the food supply for a supposed beneficial purpose are obviously the most amenable to regulation. They can be specifically banned or their concentrations limited once a harmful effect is established. However, as the number of such legal additives continues to increase, controversy often surrounds new additions to the list for a variety of reasons. I will discuss examples of some of the controversies in the various categories of deliberate additives.

### Animal Feed Additives

Chemicals added to farm animal feed are not intended to appear even in trace form in the human diet. I have previously discussed the frightening case of feeding neurological tissue to cattle and the ensuing Mad Cow Disease, and possible human nv CJD. Examples of other controversial chemical food additives to animal feed are as follows:

*Diethylstilbestrol (DES)*

DES was used extensively in cattle feed in the 1960s to hasten the fattening of beef cows so they could be ready for slaughter sooner and at a lower cost. As such, the additive was of considerable economic importance to the beef industry. DES was also used as a human prescription drug intended to prevent miscarriages in pregnant women. Eventually, human epidemiological studies implicated DES as a cause of vaginal and other cancers in the offspring of women who had taken the drug. Rodent testing also found DES to be carcinogenic to mice in doses as low as six ppb in their feed. The industry argued that when the hormone was withdrawn seven days before slaughter, no trace could be found in the meat produced. However, at the time, testing limitations prevented detection at ppb levels. Thus, safety could not be demonstrated, and DES became one of only two chemicals that were banned under the Delaney Clause, which had been amended in 1963 to allow its application to animal feed supplements.

*Bovine somatotropin (BST)*

Somatotropin is a natural hormone produced by dairy cows and is known to increase milk production in cows with high levels of the hormone. In the 1990s, a genetically engineered form of the hormone was developed that could be fed to cows with dramatic effect on milk production. The approval of BST as an additive was strongly opposed by opponents of biotechnology and by small dairy farmers who feared they would be economically disadvantaged by its use. The opposition claimed that BST was a cause of increased mastitis (a common streptococcal udder infection) in dairy cows. High production cows do have an increased tendency for mastitis infection. The result, the opponents claimed, would be a higher potential for the antibiotics used to treat the infection to appear in market milk.

Regulations have long been in force to withhold milk from cows being treated with antibiotics until such time as the milk is antibiotic free. The extent to which such regulations are complied with is perhaps inconsistent, but antibiotics can be detected in the milk with adverse consequences for farmers who do not comply. No human health effects have ever been demonstrated from BST. The genetically engineered hormone has been approved as an additive, although in some states dairy products may be labeled as "hormone free" to allow concerned consumers a choice if they are opposed to genetic engineering. Antibiotics are obviously useful in treating infections in farm animals and thus economically important to the industry. However, the possible appearance of antibiotics in meat products intended for human consumption is of concern. The potential for increasing the antibiotic resistance of human microflora is serious enough for physician prescribed medications. It should not be influenced by such chemicals ingested through the diet.

## Deliberate Additives in Human Food

Chemicals are added to the human food supply for multiple purposes. The number of such deliberate additives has increased dramatically over the years, now totaling approximately 2,000 in the United States. The Delaney Clause and the GRAS List have been the major legislative protections to assure safety of those additives. Labeling requirements are in effect to inform the public of all additive ingredients, and the consumer accepts on faith that such ingredients will cause no harm.

I will review some of the major categories of food additives and some of the publicized incidents that have accompanied their introduction and use. I should emphasize that proven human health effects from these additives are minimal, and the great majority are routinely consumed without question or controversy.

## Flavorings and Flavor Enhancers

Monosodium glutamate (MSG) has been used for many years as a flavor enhancer (salt substitute) under the trade name of Accent. It is particularly popular in Oriental cooking. Most people tolerate moderate levels of MSG without ill effect. However, a small percentage of the population is sensitive, experiencing transient neurological symptoms including dizziness and a tingling in the fingers. This reaction has been called "Chinese Restaurant Syndrome." In situations like this, where individual sensitivity, rather than general toxicity, is involved, the intervention strategy usually employed is a warning label rather than general prohibition. Certainly, people known to be MSG sensitive should avoid any foods whose ingredients labels list that chemical and should ascertain the extent of MSG use in restaurant orders.

## Food Colorings

Food dyes have been used extensively to enhance appearance of products, without providing any taste, stability, or nutritive value. Colorings are often based on coal tar derivatives, which have a high potential for causing cancer in laboratory animals. Those that have been approved as food additives have passed extensive testing under specific legislation (separate from the Delaney Clause). The tests have resulted in numerous prohibitions of particular colorings over the years. In the 1970s, there was a particularly embarrassing incident for the U.S. Department of Agriculture. The purple dye in the stamp they applied to meat products to verify that it had passed a safety inspection was found to be an animal carcinogen. An orange stamp was substituted until a noncarcinogenic purple could be found.

### Preservatives

Preservatives are certainly among the most important of all food additives. They greatly prolong shelf life and are particularly important for increasing the available food supply in areas where refrigeration is not available. Many of the common preservatives, such as sodium benzoate and sodium propionate, have been used very successfully and without controversy. One preservative which has aroused controversy, however, is sodium nitrite. I have previously discussed the acute toxicity of sodium nitrite at high concentrations, when accidentally substituted for table salt. At low concentrations it has been used as a preservative in cured meat products such as bacon, hot dogs, and cold cuts. The controversy arose when it was realized that under high-heat conditions (such as frying bacon), sodium nitrite combines with the amines present from the curing process to form nitrosamines, which have been recognized as an animal carcinogen and have been associated with human esophogeal cancer in some parts of the world. In this case, the intervention has been to restrict the concentration of nitrites to a level deemed safe. Nitrosamines also appear as a process byproduct in other consumer products such as wine and beer.

Another preservative that has been controversial is sodium sulfite, which has been used extensively in ever popular restaurant salad bars to prevent wilting of lettuce. Sodium sulfite, like MSG, provokes a sensitivity reaction in a small percentage of people. For sodium sulfite the reaction can be quite severe, including anaphylactic shock. Again, because there is no general toxicity associated with the chemical, the intervention strategy is to require warnings and to educate those who are sensitive to seek assurances that foods consumed are sulfite free.

### Artificial Sweeteners

Perhaps the most highly publicized and controversial of all food additives are the artificial sweeteners. Going back to the 1960s, there were two equally reliable calorie-free sugar substitutes legally added to the food supply. These products—cyclamates and saccharin—were believed to be important for diabetic individuals needing to restrict sugar intake and for calorie-conscious dieters. In 1969, cyclamates were banned under the Delaney Clause when rodent testing indicated potential carcinogenicity. Because saccharin was still available, the banning of cyclamates did not provoke much opposition.

However, in the late 1970s, a Canadian study indicated that saccharin also produced bladder tumors in some rats. That finding set off a major lobbying campaign by the soft drink industry, which now relied on saccharin, to prevent FDA from invoking the Delaney Clause. Much of their argument centered on the relatively high dose used in rodent testing and the possibility that a contaminant (ortho-toluenesulphonomide), not saccharin itself, was responsible for the rodent tumors. Much was made of the fact that a human would have to consume some 800 cans of diet pop per day to receive a dose equivalent to that fed to the rats. In the end,

Congress prevented the FDA from banning saccharin and, instead, opted for warning labels. A case can be made that the saccharin case study represents an example of industry pressure prevailing over the scientific process invoked for chemicals with less political clout. The other viewpoint is that a ban would have represented an overreaction based on unreasonable extrapolation of animal studies to human reality.

In any event, the controversy became moot a few years later when a new artificial sweetener became available. Aspartame (trade name Nutra Sweet) was approved for dry foods in 1981 and for diet soft drinks in 1983. That additive has remained on the market into the late nineties for cold food applications only. Aspartame produces methyl alcohol when heated. A 1996 report that aspartame might be associated with brain tumors has been mostly dismissed by the scientific community. Meanwhile, several new sweeteners have been approved. Sunnett is a heat-stable product which can be used in baked goods, and Acesulfane-K can be blended with aspartame to improve shelf life. In 1998 still another sweetener called Sucralose was approved. This chemical, ironically, is made from sugar, but is 600 times sweeter and contains no calories. Thus, the future of the diet pop industry appears secure for the twenty-first century.

One other note should be added to the artificial sweetener story. Still lacking is any credible evidence that the enormous growth in artificially sweetened products has had any effect on weight loss in diet-conscious Americans.

**Fat Substitutes**

In 1996 the FDA approved the first artificial fat as a food additive. Olestra (trade name Olean) is marketed by Procter & Gamble as a cooking oil and in snack foods. The product is not without controversy. Because it is based on very large molecules designed to pass through the digestive tract without being absorbed the way real fat is, Olestra is suspected of causing diarrhea when consumed in large quantities. In addition, vitamins and other beneficial nutrients pass through the digestive tract without being absorbed. Several other artificial fats are under development as this is written. One of those is based on oats and may actually prove to be a flavorful and healthy dietary addition.

**Nutritional Supplements**

A few other additives have come in for criticism and scrutiny. Vitamin supplements are clearly a big business and are believed to be important nutritionally by many people. Others question the need for such supplements, emphasizing a healthy balanced diet instead. Vitamin C is promoted by some as a preventative for the common cold, a theory disputed by others. Possible adverse effects of vitamin overdoses are also cited. Vitamin A in high doses may produce a condition which mimics a brain tumor, while vitamin D may result in calcification of soft tissue, particularly affecting people with arthritis.

A few other health effects of nutritional supplements have been noted. One example is the presence of amygdalin (cyanide) in apricot kernels sold as a nutritional supplement. Another problem has been reported with tryptophan, an amino acid appearing naturally in some foods, notably turkey meat. Tryptophan has been produced artificially by genetic engineering and marketed as a treatment for depression and insomnia. In 1989 it was epidemiologically associated with an outbreak of eosinophilia-myalgia syndrome (EMS). That outbreak was thought to be the result of a contaminant introduced through a manufacturing error by a Japanese producer of the genetically engineered tryptophan. Finally, in 1996, consumption of an herbal supplement called ephedrine was blamed for an outbreak of some 500 illnesses and 17 deaths. This supplement was neither tested by nor regulated by the FDA due to a loophole in the regulations.

# Section V

# INTEGRATED PEST MANAGEMENT

# Chapter 14

# PUBLIC HEALTH SIGNIFICANCE OF RODENTS AND INSECTS

On a worldwide basis, zoonotic and other infectious diseases transmitted to humans through the various machinations of arthropods and other animal species still represent the most important of all public health problems. The ability of these infectious agents to jump across species barriers to infect humans presents a formidable challenge to our survival instincts—not to mention our public health infrastructure. In the United States, we are somewhat less threatened by vector-borne diseases, but there are still significant examples of their occurrence. When global trade and transportation trends are considered, we cannot become complacent that we will remain isolated from problems anywhere on the planet. A good example is the importation of the tiger mosquito (*Aedes albopictus*), previously not found here, in used tires imported from Asia. This mosquito has already been shown to be capable of transmitting arboviral encephalitis in the United States. The potential for ticks and other ectoparasites to arrive on imported livestock and the potential for infected humans to travel around the world—in less time than the incubation period of many infectious diseases—signals the importance of constant surveillance to counter these threats.

In this chapter I will discuss a variety of the most important zoonotic and arthropod-borne diseases demanding our attention at the turn of the millennium.

## DISEASES SPREAD BY DOMESTIC AND FERAL RODENTS

### Plague

Plague is one of the few human diseases in recorded history to cause a downward blip in the exponential growth of the human population. The so-called "Black Death" which ravaged Europe in the fourteenth century is said to have killed one-third of the population during its relentless march across the continent. Subsequent pandemics of plague continued to erupt periodically until the early twentieth century. The physical, economic, and psychological devastation caused by plague in earlier times, before the germ theory of disease was accepted, can only be imagined. We now know the chain of transmission and can cure diagnosed cases with antibiotics. However, in the Dark Ages the cause was unknown. The harbinger of plague was the sight of dead rats littering the streets of the town. Before modern sanitation, rats were very abundant, but, being nocturnal, were rarely seen in daytime. We postulate that the Norway rats (now known to be the reservoir of plague) followed the human trade routes, spreading the epidemic from town to town along the way.

The disease is kept going among the rodents by an ectoparasite (the oriental rat flea, *Xenopsylla cheopis*). The infectious agent is the bacterium *Yersinia pestis*. This pathogen is amazingly inefficient in that it sickens not only its reservoir, but also the flea vector. Humans are infected incidentally when the host rat dies and the flea must seek a secondary host. The flea regurgitates the organism into the bite wound on the human skin. The result is the bubonic form of plague characterized by lymph node involvement. Secondary lung infection can result in person-to-person droplet transmission of the more dangerous pneumonic plague through coughing and sneezing.

Although urban plague has been absent in the U.S. since the 1920s, a rural sylvatic form still exists in feral rodents in the Southwest. A handful of cases are reported every year in persons having contact with wild rodents in that area. In 1994 there was a classic outbreak of urban plague in the city of Surat in India. The outbreak was marked particularly by the pneumonic form of the disease. The outbreak was attributed to a very high rat population resulting from heavy rains and unsanitary conditions in the city, serving as a reminder that continued attention to urban rat infestations is still needed.

**Endemic (Murine) Typhus**

This disease, which still occurs at very low endemic levels in the Southeastern United States, is caused by a rickettsial agent (*Rickettsia typhi*). The epidemiologic chain is very similar to bubonic plague—a rat reservoir, flea vector, and accidental human host. Recent information suggests that in addition to domestic rats, feral cats, dogs, and opossums may harbor the agent.

**Leptospirosis**

A very different mechanism is involved in the transmission of leptospirosis. Rats are only one of numerous animals harboring the leptospirosis agent (various serovars of the spirochete *Leptospira interrogans*). Cattle, dogs, and raccoons are also thought to be reservoirs. Transmission to humans occurs primarily by bathing in water contaminated by the urine of host animals. The portal of entry involves contact with abraded skin or mucous membranes. In 1995 there was an outbreak of leptospirosis in Nicaragua, apparently precipitated by an increased rat population following heavy rains. Additional cases were reported following hurricane induced flooding in 1998. In the United States, leptospirosis incidence is generally in the annual range of 30 to 60 cases. However, there is renewed concern in rundown urban areas where a high percentage of rats are positive for leptospirosis, and human cases have been reported from cuts sustained from alley trash containing rat urine soaked debris.

### Ratbite Fever (Haverhill fever)

This bacterial disease caused by *Streptobacillus moniliformis* is transmitted directly by rat bites and occasionally by contaminated food or water in rat-infested buildings. Ratbite fever was mentioned in Chapter 8 in the discussion of substandard housing problems, which include direct attacks by rats on vulnerable humans.

### Hantavirus Pulmonary Syndrome (HPS)

HPS is a classic emerging infectious disease that surfaced in 1993 in the Four Corners area of the Southwestern United States. The causative agent was quickly identified as a new hantavirus since named Sin Nombre virus (or "virus without a name"). The illness differs from previously identified hantavirus diseases in that it causes pulmonary rather than renal symptoms. Previous outbreaks had occurred in Scandinavia and in Korea among American troops during the Korean War. It is believed that more than 100,000 cases of hantavirus disease occur in Asia every year. The Four Corners outbreak was particularly scary in that it victimized previously healthy young adults and resulted in about a 50 percent fatality rate within a few days.

The subsequently identified chain of infection involves primarily a deer mouse reservoir. Humans are infected by inhalation of aerosols generated by stirring up dried excreta from the mice. The 1993 outbreak was associated with a very heavy population of deer mice that year nourished by a bumper crop of piñon nuts following heavy winter rains. As of 1998, more than 160 cases from more than 20 states had been identified in the United States, and more than 400 cases in the western hemisphere, reflecting the broad geographic range of the rodent vector. Similar conditions in 1998 may have precipitated additional cases. Outbreaks in South America may involve person-to-person transmission in addition to the aerosol route. Intervention is difficult due to the varied habitat of the deer mouse. Education of the public to avoid aerosolizing rodent excreta in closed spaces is a necessary intervention step.

### Arenaviruses and Filoviruses

Among the many emerging exotic viruses in tropical areas, a few have been particularly publicized. Lassa fever (an arenavirus) is associated with wild rodents in parts of Africa where the disease is endemic. Like HPS, transmission is via aerosolized rodent excreta or direct contact with the infected rodent.

Ebola-Marburg viruses are in the filovirus group. They were popularized through a book by Richard Preston called *The Hot Zone,* which described an incident in a laboratory in Reston, Virginia, in which nonhuman primates were infected with the virus. Despite a considerable scare, that strain (*Ebola reston*) did not prove pathogenic to humans, although 80 percent of the workers exposed to the monkeys tested positive for the virus. Still unknown is the source of the virus in the monkeys, which were purportedly imported from the Philippines, where the virus has never been detected. Outbreaks in Africa have infected humans on a number of occasions, most notably the episode in Kikwit, Zaire, in 1995. Although wild rodents or bats are suspected reservoirs of these viruses, no reservoir had been identified as of 1998 despite extensive testing of hundreds of species following the Kikwit outbreak. The search for a possible reservoir has focused on monkeys inhabiting the jungle canopy and seldom emerging at ground level. Transmission to humans appears to be primarily through direct contact with skin and sweat of patients. Direct caregivers, including family members and healthcare workers, are most at risk. It is suspected that the original (index) case may have acquired the virus by ingestion of chimpanzee meat or contact with chimpanzee blood during the butchering process.

There is not complete agreement on the potential for global spread of exotic viruses such as ebola. The victims seem to be immediate caregivers infected uniformly through close contact. In the absence of aerosol spread, wide dissemination is less likely. Also, while victims incubating the disease could theoretically travel long distances before showing symptoms, it is argued that the likely victims are extremely unlikely to be persons engaged in intercontinental travel. Nonetheless, these exotic viruses are being recognized more frequently. They appear to involve human infringement on previously uninhabited areas and encounters with previously unknown vectors. The rapidly shrinking rain forests harbor more than 50 percent of all living species. There are currently many unknowns in the potential for wider dissemination, and close surveillance is seen as the key to future limitation of problems related to emerging infections of all kinds.

**Diseases Associated with Domestic Mice**

Even more common than human contact with domestic rodents is human contact with domestic mice. Like domestic rats, mice are known to harbor common human intestinal pathogens such as salmonella. However, only a few diseases have been shown definitively to be transmissible from domestic mice to humans. They include rickettsial pox and lymphocytic choreomeningitis (LCM). Rickettsial pox is a relatively mild and uncommon illness of humans, caused by the rickettsial agent (*Rickettsia akari*). It is actually transmitted to humans by a mite—an ectoparasite of the house mouse that was first identified in 1949 in a tenement building in New York City. The same agent has been identified in the former Soviet Union, where the reservoir is thought to be a domestic rat. LCM is an arenavirus with its primary reservoir in domestic mice. The mice transmit the virus to their offspring. Transmission to humans is primarily by contact with urine, feces, or saliva of the rodents. Humans have also been infected when the virus has been transmitted from domestic mice to pet hamsters or to laboratory rodents. The incidence of LCM in humans is not known but is believed to be underreported.

## DISEASES TRANSMITTED BY ARTHROPODS

Arthropods (joint-footed creatures with exoskeletons) of public health importance include, primarily, insects (with six legs and three body parts) and arachnids (with eight legs and two body parts). Of the several million insect species which inhabit the planet, many as yet to be identified, only a few are known vectors of human disease. Of those, by far the most significant to human health are the mosquitos.

### Mosquito-borne Diseases

*Malaria*

More than 100 million malaria cases and nearly 2 million deaths from malaria occur annually throughout the world. More than 2 billion people live in tropical areas endemic for malaria. The vector responsible for transmitting this disease is the anophelene mosquito. These vectors are also found in temperate zones—including the United States, where malaria was once endemic and occasional cases still occur. In fact, about 1,500 cases of malaria are reported annually in the United States, almost all in travelers from endemic countries. The few cases in persons who have not traveled abroad are referred to as autochthonous; these have been limited to one or two per year because the number of infected humans who serve as the reservoir is extremely limited and the climate does not favor the life cycle of the parasite. The chain of transmission is as follows:

1) *Agent.* The agent is a protozoan called a plasmodium. There are four species responsible for human malaria (*Plasmodium vivax, P. malariae, P. falciparum,* and *P. ovale*). *P. falciparum* is the most serious, but all can cause lingering symptoms which can debilitate large numbers of people for many years, contributing to economic stagnation in developing countries.

2) *Reservoir.* Humans are the only important reservoir of malaria. Nonhuman primates are infected with other plasmodium species but do not contribute to the human infection chain.

3) *Transmission.* Malaria is a typical example of biological transmission in which the parasite must undergo part of its development cycle in the mosquito vector. The parasites at the sexual stage (gametocytes) circulating in the blood of the human host are removed by the bite of a female anophelene mosquito. The gam-

etocytes infect the mosquito, develop into sporozoites (which reach the salivary gland of the mosquito), and are injected into the new human host when the mosquito takes another blood meal. In the human host the parasite completes its life cycle as the sporozoites become schizonts and then merozoites, which eventually reach the bloodstream, where they mature into male or female gametocytes. Because malaria is a blood-borne disease, there were initial fears that human AIDS might be transmitted by mosquitos. However, the AIDS virus does not infect mosquitos, and the titer of virus particles which might be transmitted via a mosquito bite is far too low to initiate infection.

4) *Host.* Susceptibility is almost universal except for a few genetically resistant individuals. Travelers to endemic areas are advised to take preventive drugs such as chloroqine and mefloquine. Environmental intervention strategies will be discussed in Chapter 15.

*Yellow fever*

Yellow fever is caused by a viral agent in the flavivirus family. It is spread by the bite of the *Aedes aegypti* mosquito and was an important deterrent to early attempts to build the Panama Canal. Humans are the principal reservoir, although transovarial transmission in the mosquito is suspected to play a role, and monkeys and marsupials may also be carriers. Vector control programs were initially responsible for prevention of the disease, but an effective vaccine was eventually developed. A resurgence of yellow fever in Africa in the 1980s and 1990s has raised concern that this disease may be reemerging. Urban areas in many parts of the world are susceptible because of the wide distribution of the *Aedes aegypti* vector.

*Dengue fever*

Similar to yellow fever, dengue is a viral disease also transmitted by the bite of the *Aedes aegypti* mosquito. It is considered one of the more important emerging infectious diseases as its incidence has been increasing rapidly in endemic areas of the Caribbean and in Southeast Asia. The hemorrhagic form of the disease, which occurs primarily in Southeast Asia, can be very serious. No vaccine is currently available; prevention depends primarily on vector control.

*Arthropod-borne viral encephalitis*

There are numerous varieties of viral mosquito-borne encephalitis. Unlike the tropical endemicity of the previously discussed diseases, many encephalitis cases occur in temperate climates, including the United States. Some of the varieties include Eastern and Western Equine, Japanese, St. Louis, La Crosse, California, Murray Valley, Jamestown Canyon, and Rocio Encephalitis. This group is perplexing in that the viruses can be transmitted by a variety of mosquitos, and for the most part specific reservoirs have not been identified. Various wild birds, small mammals, and the eggs, larvae, and adult forms of the mosquitos themselves are all suspected reservoirs. Horses (in the equine varieties) and humans are both dead end victims of the disease; titers in large animals do not build up to the point where transmission to other animals is possible. In the United States, important mosquito vectors that have been identified include *Culiseta melanura, Culex tarsalis,* and *Aedes triseriatus* (the tree hole mosquito).

*Other mosquito-borne diseases*

Several other mosquito-transmitted diseases are worthy of mention. Filariasis (also called elephantiasis), for example, is a helminthic disease caused by a nematode (*Wuchereria bancrofti* and several others) with close to 100 million cases reported annually from endemic areas with warm, humid climates in Africa, Asia, and South America. It has a human reservoir and is spread by numerous different mosquitos. West Nile fever, another example, is of some concern following an unusual urban outbreak in Bucharest, Romania, affecting some 500 people in 1996. In fact, Benenson (Control of Communicable Disease Manual, APHA, 1995) lists more than 60 different viruses known to be transmitted by mosquitos. The ongoing encroachment of humans into forest and jungle areas, previously uninhabited, raises the potential for continuing emergence of infectious diseases with unpredictable consequences.

*Burkitt lymphoma*

Before leaving the subject of mosquito-borne disease, some mention should be made of Burkitt lymphoma, a monoclonal tumor of B cells affecting mostly young children in tropical Africa and Papua New Guinea. About 97 percent of the cases have been associated with Epstein-Barr virus (EBV), the same virus that causes mononucleosis in the United States. The geographic distribution is consistent with high malaria endemicity. Thus, a possible mosquito link is postulated but by no means proven.

**Other Insect Vectors of Human Disease**

While mosquitos are far and away the most important insect vectors of human diseases, many other insects have also been implicated. I have previously discussed the role of the oriental rat flea (*Xenopsylla cheopis*) in plague and endemic typhus. That is another classic example of biological transmission. Similarly, the human body louse (*Pediculus humani corporis*) is the vector of epidemic louse-borne typhus fever. That disease has been important historically, reportedly killing more soldiers during the Napoleonic Wars than the battlefield itself. The disease has a human reservoir (although there is also a reported zoonotic reservoir in flying squirrels). Epidemic typhus has occurred in military and civilian refugee populations in times of sanitation breakdown when bathing and laundry facilities were unavailable. Control obviously includes provision of such sanitary facilities, but has also been accomplished in their absence by insecticide dusting of human refugee populations. The list of other insects includes the cone-nosed bug transmitting Chagas disease; the blackfly as a vector of onchocerciasis (river blindness); tsetse flies, which transmit African trypanosomiasis (African sleeping sickness); and sandflies, which transmit a group of phleboviruses causing sandfly fever.

### Mechanical Transmission of Human Diseases

No discussion of the public health importance of insects would be complete without mention of houseflies and cockroaches. Infestations of those insects in human environments, particularly food preparation areas, is virtually synonymous with sanitation breakdown. Neither the fly nor the cockroach has been proven to be a biological vector of human disease. However, both are filth breeders which inhabit the kitchens of human civilization. As such, they are looked upon as potential physical carriers of infectious enteric agents from fecal matter to food. Cockroaches have also been identified as a source of human allergens and are suspected of playing a role in childhood asthma. Thus, prevention of fly and roach infestations is an integral part of food sanitation practice. Such prevention will be discussed in Chapter 15 under integrated pest management practices.

## ARACHNIDS AS VECTORS OF HUMAN DISEASE

Just as mosquitos are the most important insect vector of human disease, ticks are by far the most important arachnid vector. Unlike mosquitos, which wreak most of their carnage in tropical climates while serving mainly as an annoyance elsewhere, the ticks are abundant and actively transmitting diseases throughout North America and other temperate regions.

### Tick-borne Diseases

#### *Lyme disease*

Lyme disease accounts for approximately 75 percent of all confirmed vector-borne disease in the United States in the late 1990s. The disease peaks at the height of the tick season in late spring. Early symptoms are variable, complicating early diagnosis. The illness often lasts for months and may become chronic. Cardiac involvement and rheumatic sequellae are common. Since it became nationally reportable in 1989 the number of reported cases has increased to approximately 15,000 per year with about 80 percent of those from southern New England and the mid-Atlantic states. There is also a focus in western Wisconsin and eastern Minnesota and another in northern California and Oregon. However, increased surveillance resulted in reported cases from some 47 states by 1998.

Lyme disease is caused by a bacterial spirochete (*Borellia burgdorferi*), which was not identified until 1982. The reservoir is primarily in small rodents, particularly deer mice of the *Peromyscus* species. The vector is the ixodid tick (*Ixodes scapularis* in the eastern and midwestern United States, *I. pacificus* in the western states and several other ixodid species in Europe and Asia). Transstadial (from one stage of the life cycle to another) transmission in the ticks has been identified. The deer population serves as an important maintenance host for the ticks. Larval ticks feed on the small rodent hosts while adults feed on the deer. Prolonged attachment of the tick (about 24 hours) appears to be necessary for transmission to humans.

Intervention is complicated by the difficulty of targeting the vector ticks in their wideranging woodland habitat. Personal protective clothing, tick repellents, and self-examination during daily bathing or showering during tick season are emphasized. Ixodid ticks are very small relative to the more common and may be difficult to detect on the human body. Reporting and diagnosis of Lyme disease are somewhat controversial. Serologic tests are poorly standardized. There is some thought that the disease is overdiagnosed, leading to unnecessary, expensive antibiotic regimens.

*Human granulocytic ehrlichosis (HGE)*

This recently identified disease is caused by a bacterium (*Ehrlichia phogocytophila*) with an unknown reservoir. It is apparently transmitted to humans by the same ixodid ticks which are vectors of Lyme disease, but may also be spread by the common dog tick. HGE apparently can coexist with Lyme disease in the same patient, further complicating diagnosis.

*Rocky Mountain spotted fever*

Approximately 500 cases of Rocky Mountain spotted fever are diagnosed in the United States each year. About 50 percent of those are in the southern Atlantic states. It is a rickettsial disease (the agent is *Rickettsia rickettsii*). It is transmitted through the bite of the common dog tick (*Dermacenter variabilis*), although wood ticks and Lone Star ticks have also been implicated in some locations. The ticks also serve as the reservoir through transovarial and transstadial passage.

*Q fever (Query fever)*

This disease is caused by a rickettsial agent (*Coxiella burnetti*). Unlike the other diseases discussed in this section, ticks serve as one of many reservoirs for the agent, along with sheep, cattle, goats, dogs, cats, and some birds. The ticks probably play a major role in maintaining the natural cycle in the other animal reservoirs but not as a direct vector in human transmission. Q fever is primarily transmitted to humans via the airborne route in particles disseminated from sheep placentas. The infectious dose for humans is very low, with infection possibly being initiated by a single organism. A community outbreak that occurred in Rollhausen, Germany, in 1996 was traced to long-distance aerosol transmission from a sheep farm more than a mile away. The Q fever agent has been found in raw milk, but milk has not been confirmed as a vehicle of transmission to humans. Because the agent is relatively heat resistant, batch pasteurization requirements (145°F for 30 minutes) are actually based on the heat resistance of the Q fever organism.

### Other Arachnids as Human Disease Vectors

In addition to ticks, another arachnid which has been implicated in transmission of human disease is the common mite (*Liponyssoides sanguineus*). This arachnid, as an ectoparasite of house mice, was previously mentioned for its role in rickettsial pox. Similarly, scrub typhus, another rickettsial disease, is spread to humans through the bite of a larval trombiculid mite (*Leptotrombidium akamusbi*). The mites also serve as the reservoir for the rickettsial agent.

## SNAILS AS VECTORS OF HUMAN DISEASE

Schistosomiasis is an extremely important disease with some 200 million cases reported annually worldwide. It is endemic in areas of Africa, South America, and Asia that are lacking in human wastewater carriage and treatment infrastructure. The causative agent is a helminthic trematode, the blood fluke *Schistosoma*. The life cycle and chain of transmission are complex. Humans are the principal reservoir, although other domestic animals may be infected. The blood fluke eggs are excreted in human urine and feces and contaminate water in lakes or streams. The eggs hatch to produce miracidia (the larval form). The snail is the intermediate host, where the miracidia grow into free-swimming larval cercariae. When these are released, they penetrate the unbroken skin of people wading in the water. The larvae mature in the human host to produce the eggs which are excreted to continue the cycle. Control is very difficult in the absence of wastewater facilities because the snail vector has proved to be an elusive target for integrated pest management efforts.

## REFERENCE

The previously mentioned APHA publication on control of communicable diseases, updated at five year intervals, is a particularly useful reference for this chapter:

Benensen, A. S. 1995. *Control of Communicable Diseases Manual* (16th Ed.) A.P.H.A. Washington, D.C.

# Chapter 15

# APPLICATIONS OF INTEGRATED PEST MANAGEMENT

The concept of integrated pest management (IPM) has become widely accepted not only as applied to the vermin of public health importance discussed in Chapter 14, but across the entire field of agriculture. In essence, the concept is to decrease reliance on potentially dangerous chemical pesticides by integrating chemical applications into a more specifically targeted approach to pest control using a wide variety of methods. After learning the life cycle and ecology of the target pest and identifying the weakest link in that life cycle for attack, intervention is then achieved by combining judicious pesticide use with engineering, sanitation, and natural or biological control methods designed to avoid harm to nontarget species. IPM can best be illustrated by citing examples pertinent to the pests of public health importance.

## THE DOMESTIC RAT EXAMPLE

The Norway rat, a very adaptable and persistent pest, has learned to live on the periphery of human habitats, making its living off of the abundance and carelessness of human civilization. Rats are omnivorous, aggressive, nocturnal, and prolific animals. They produce up to 20 young per litter with a short gestation period, enabling them to produce as many as four litters per year. They are quite athletic and can jump, swim, and scurry very efficiently. They also must gnaw to wear down their incisors, and in doing so can gnaw through wood, wires, and thin aluminum. They are burrowing animals, and will nest in banks near railroad tracks or in piles of debris indoors or out. They will stay within about 150 feet of their nests as long as food and water are available, following pathways near walls (they do not see well) to the food supply. They leave trails of droppings and grease marks from their whiskers, helpful in tracing their routes for control purposes.

In urban settings, rat infestations may be in buildings, in banks along railroad tracks, or in sewers, where they take advantage of the generous food supply provided by garbage grinders. The IPM approach differs for these different habitats. For example, potent rodenticides can be used safely in sewers because children and domestic pets are excluded there, but not in tenements or restaurants. For indoor environments, the only permanent control must rely on engineering and sanitation methods to deny rats access to food, water, and harborage. Rat-proofing buildings utilizes knowledge of rat ecology. Thick aluminum screens on windows and concrete basement floors to preclude burrowing into buildings are necessary in threatened buildings. Scrupulous cleanliness in kitchen areas (also needed for cockroach control) and storing food in secured areas and in gnaw-proof containers are also part of this approach.

Chemical rodenticides are part of the IPM approach as well. In inhabited buildings, only rodenticides with built-in safety features should be used. Red Squill is an example of such a chemical. It is a natural emetic, effective against rats because they are physiologically incapable of regurgitation, but safe for humans or domestic pets because they will regurgitate it if accidentally ingested. However, Red Squill is limited in effectiveness because rats will not return to the bait if they get a sublethal dose.

The most successful class of rodenticides has been the anticoagulants (such as Warfarin). Unlike Red Squill, rats do not associate Warfarin with the progressive weakness caused by internal bleeding that they experience from the anticoagulant, so multiple exposures are achieved. This serves as a safety feature in that accidental single exposures will not be harmful to children or pets. There is also an antidote (vitamin K) available. Rats have shown some resistance to Warfarin (perhaps induced by the vitamin K in their diets of human food scraps) and newer anticoagulants such as coumarin and brodifacoum have been developed. Another alternative to the anticoagulants is cholecalciferol (Quintox or Rampage), which serves to interfere with calcium metabolism in the rat.

For outdoor situations where the rats are entrenched in burrows (for instance, under railroad tracks), carbon monoxide (CO) gas can be used, simply by attaching a hose to a vehicle exhaust system and pumping the CO-laden exhaust into the burrow opening. Identifying and sealing alternative burrow exits may be necessary with this approach. Finally, for sewer infestations, more potent rodenticides such as sodium fluoracetate may be used as long as human and pet access is prevented. The more potent rodenticides should be used only by licensed exterminators.

Natural or biological methods are also part of IPM for domestic rats. Natural predators such as dogs and cats are quite effective against both rats and mice. The ferret and the mongoose are also rodent predators but have been utilized less frequently. The barn cat is employed primarily to keep rodent populations in check. Other biological controls for domestic rats have been attempted, such as seeking a bacterial pathogen specific to rats. However, rodent physiology is apparently too close to that of humans; rodent pathogens such as *Salmonella* and *Yersinia* are also pathogenic to humans. Antifertility drugs are another possibility, which thus far has proven too costly.

### The Importance of the Time Sequence

The IPM approach requires thinking out all potential consequences of control actions. In domestic rat control, for instance, rats should not be killed with rodenticides if they are carriers of pathogens such as those of plague or endemic typhus, transmissible to humans by fleas. Live trapping to test for such pathogens should be step one in endemic areas. If rats test positive for pathogens, then flea control should be accomplished before baiting with a rodenticide. Dusting identified rat runways with an effective flea powder is the next step—another "rat fact" is that they routinely preen themselves, thus distributing the dusted powder over their bodies to assist in flea control.

Knowledge of the runway paths also assists in the next step, which is baiting with one of the "safe" rodenticides to reduce the population. This should be done before rat-proofing of the building or final sanitation cleanup to prevent rats from migrating elsewhere in search of food or harborage. Traps may be needed to eliminate the last elusive invaders. Finally, rat-proofing and sanitation cleanup are the necessary last steps for permanent control.

## IPM FOR MOSQUITO CONTROL (WITH VARIATIONS FOR OTHER INSECTS)

The mosquito provides an excellent example of the necessity for IPM. The extreme importance of mosquito-borne diseases such as yellow fever at the start of the twentieth century, before potent insecticides were available, prompted the development of IPM methods long before the concept became fashionable or specifically designated as such.

The mosquito life cycle is an example of what is called complete metamorphosis. Mosquito eggs are laid in places subject to subsequent flooding. The eggs hatch in water producing the larvae which represent the growth and feeding stage of the cycle. Mosquito larvae are called wrigglers, which have no resemblance to the adult. The larvae are clearly the most vulnerable stage of development and thus the target of most control efforts. Larvae eventually reach full size whereupon they pupate. The pupa is the stage in which the larva undergoes metamorphosis into the flying adult.

Different species lay the eggs in different places, complicating control approaches. There are more than 200 species of mosquito in the United States. One of the most important is the *Aedes* genus, including *A. aegypti*, the vector of yellow fever and dengue; *A. trisiterus,* the "tree hole" mosquito, which is an important vector of arboviral encephalitis in temperate climates; and *A. albopictus*, the "tiger" mosquito, imported into the U.S. via old tires in 1988, and also a potential vector of dengue and encephalitis. All of the *Aedes* mosquitos lay eggs in odd places, such as tree stumps, birdbaths, and tin cans (or old tires). They thus provide an elusive target for larval control compared to those species which choose more predictable settings, such as lowlands or roadside ditches, which flood during the rainy season.

Heavy rains invariably make the control task more difficult by increasing the temporary wetlands and often leading to enormous hatches of the pest.

## Engineering Approaches to Mosquito Control

Screens are the most effective approach to preventing mosquito entry into dwellings, and in endemic malaria areas, netting over beds is routinely employed. Sanitation cleanup is an important ingredient for control of the *Aedes* species. Engineering methods have also been utilized in attempts to modify breeding sites to make them unsuitable for the mosquito life cycle. Chemical treatment of soil to make it more porous and to prevent standing water is one example. Cutting down vegetation along stream banks to prevent stagnation, which favors larval development, is another.

An innovative engineering approach called the "oasis" method was employed in South America in the 1930s. This approach actually combined engineering with a natural control method and is mentioned to illustrate the integration aspects of IPM. A deep pool was excavated in a low-lying area, and trenches or canals were dug radiating outward from the pool into the surrounding lowlands. The pool was stocked with a predator fish called gambusia. During the rainy season, the lowland would flood and the fish would swim out to feast on the hatching mosquito larvae. When the water receded during the dry season, the fish could return through the canal system to the permanent pool or "oasis" to survive and thrive until the next cycle of flooding.

## Chemical Pesticides and IPM

As has been previously discussed, the use of chemical insecticides has become one of the most contentious of environmental health issues. These chemicals have had a significant impact on control of vector-borne diseases and the production of high-yield agricultural output. However, the suspected hazards to human health and to wildlife have tempered the initial enthusiasm and led to current efforts to utilize these pesticides in a more responsible manner consistent with the goals of IPM. It is also important to note that a distinction must sometimes be made between long-term human health effects from pesticides used in agricultural production (where the great majority of pesticide use is concentrated) and short-term use to control infectious disease outbreaks which could threaten lives in the immediate present. Thus, some pesticides, banned or restricted in the United States in agricultural use, may still be used in developing countries for control of disease vectors of immediate importance to public health.

*Chlorinated hydrocarbons*

The greatest ecological problem with insecticides is that of long half-life with prolonged persistence in the environment. The chlorinated hydrocarbons represented by DDT (dichlorodiphenyltrichloroethane) are the best example of that problem. The half-life of DDT has been estimated to be in the range of two to four years. First formulated in 1874 in Germany by a chemistry student, DDT and its insecticidal properties were recognized in 1939 by a Swiss chemist named Paul Muller, who was subsequently awarded the Nobel prize for the discovery.

DDT was effective against a broad spectrum of insects, and its low acute toxicity allowed it to be used as a powder on human refugee populations in World War II for control of body lice. It could also be formulated as an emulsion to float on mosquito-breeding ponds to kill larvae that surfaced to breathe and feed. Its persistence allowed it to be used very economically as a single application that lasted for months. As an aerosol it could be sprayed from crop-duster airplanes for treatment of vast acreages of field crops.

No wonder DDT was seen as a significant breakthrough with enormous agricultural and public health implications. The problem was that 75 percent of the chemical never got to the target crops. It was instead washed into streams and eventually into oceans, or aerosolized for long-distance transport. It eventually made its way to all regions of the planet, and into the food chain where biomagnification in some species increased its concentration. It appeared in the fatty tissue of many species, including penguins in the Antarctic and in human mothers' milk. It was ultimately suspected of interfering with the reproduction of many wild birds by thinning the eggshells in some species. The potential hazards were publicized by Rachel Carson's book *The Silent Spring* in 1962.

In spite of the usual scientific uncertainties and obfuscation regarding real versus perceived hazards to humans and other species, DDT was effectively banned in the United States in 1972. Numerous other chlorinated hydrocarbons with properties similar to DDT were also formulated. Many, including chlordane, aldrin, dieldrin, and heptachlor have also been greatly restricted or banned. A few, including lindane and toxophene, continue to be used.

*Natural derivative insecticides*

Many other classes of insecticides have also been formulated. Of interest to IPM has been the natural derivative group. They are sometimes called "knockdown" insecticides because they are immediately effective but do not persist in the environment. They also have very low human toxicity. Pyrethrum, a derivative of the chrysanthemum plant, is typical of that group, and many variations based on similar chemistry have been artificially formulated. The oil of the neem tree, a native of India, has been recognized as having insecticidal properties and is being developed as a natural insecticide. Similarly, oil of citronella is a natural insect repellent.

*Organophosphates*

The restrictions placed on DDT and other chlorinated hydrocarbons opened the door for an increased role for the organic phosphates. This group is also effective against a broad spectrum of insects but has a much shorter half-life (about one month). Human acute toxicity is very variable for the organic phosphates. Parathion, widely used in agriculture, can be absorbed through the skin and is extremely toxic. Many occupational fatalities have been attributed to that chemical. On the other hand, malathion, has very low acute toxicity and has been used successfully against adult mosquitos. Abate is another organic phosphate with an acute human toxicity so low that it has been used as a mosquito larvicide in drinking water reservoirs. Thus, this group has a place in IPM but is clearly more expensive than the chlorinated hydrocarbons, and must be applied more frequently and with extreme caution.

*Carbomates*

The carbomates act as stomach poisons. The most frequently used carbomate is carbaryl (trade name Sevin). Used in many agricultural applications, it has low toxicity for humans and for wild birds. However, it is lethal unselectively to many useful insects, such as honeybees, and it reduces the insect population required by insectivorous birds.

**Insect Repellents and IPM**

Mosquito and tick repellents such as DEET are an alternative prevention strategy to the chemical pesticides. They can be particularly important for prevention of Lyme disease because of the lack of effective tick eradication methods. They are also important in endemic mosquito-borne disease areas where vector control programs are lacking.

**The Time Release Concept**

One of the most important IPM developments in the past few decades has been the use of time release methods for pesticide application. This approach is widely used in agriculture and even by home gardeners. By using time release capsules, short half-life pesticides with many fewer ecological disadvantages can be applied more economically. The pesticide is released gradually over an extended time period so that reapplication is not necessary.

## NATURAL OR BIOLOGICAL METHODS OF IPM

As IPM has become established for numerous pest control applications, there are many examples of sophisticated use of knowledge of life cycles to combat the adversary arthropods. With the growing importance of Lyme disease, for example, we are starting to see more detailed study of tick habitats. One approach is to impregnate cotton balls with a tick larvicide called permethrin. The cotton is simply piled in wooded areas for use by the deer mice as nesting material, the idea being that the mice themselves will distribute the larvicide during preening and control the larval ticks which feed on them. New studies are also pinpointing concentrations of tick populations for targeted control rather than assuming that they are distributed uniformly throughout the woods.

### Natural Predators

Many species of fish apparently feed on mosquito larvae; birds feed on adult mosquitos. The gambusia were mentioned previously in conjunction with engineering of the "oasis" as an early example of IPM. Top minnows have also been shown to be predators of mosquito larvae. However, there should be a word of caution relative to introducing new species into any ecosystem. The results may not justify the original intent. Predators usually have voracious appetites, not only for the target species but for beneficial species as well. They may eat species that are actually more specifically predators of mosquito larvae, thus actually increasing the mosquito population. Again, more detailed research should be completed to assure optimum results. Insects such as ladybugs and aphids show promise for natural control.

### Natural Hormones

Several natural hormones can be synthesized for use in IPM. Pheromones are an example of a hormone that has found application. The pheromones are sexual attractants specific for a given insect. They can be synthesized and then used to lure the insects to traps where potent insecticides can safely destroy the target insects without widespread environmental dissemination. Another example is the use of "Juvenile Hormone" for controlling mosquito larvae. The hormone is a natural substance which prevents the larva from pupating once it has reached maximum growth. Synthesized as Altosid or Methoprene, this hormone has effectively controlled mosquito populations in some applications. The cost of such innovations has been reduced as biotechnology has improved. Another innovative example emerging in the late 1990s is the use of imidocloprid on cockroaches. It has been determined that roaches, which lack an effective immune system, ward off pathogens by constant grooming to prevent the pathogens from penetrating. Imidocloprid inhibits grooming, therefore allowing a pathogenic fungus to penetrate and kill the roach.

### Microbiological Agents

While the search for pathogens specific for rodents but harmless to humans has proved elusive, a number of insect pathogens have been used successfully. Foremost among those is *Bacillus thuringiensis* (Bt), a spore-forming bacterium naturally occurring in soil. It was noted as early as the 1930s that these organisms prevented Japanese beetle devastation of garden plants. The toxin is deadly to numerous insects. The toxin gene has now been isolated and can be bioengineered for greater application. For example, for use against mosquito larvae, the gene is transferred into an aquatic bacterium (*Caulibacter crescentus*) that survives well in mosquito habitats unfriendly to the soil-dwelling Bt.

The Bt gene has also been incorporated by biotechnology directly into cotton seeds (trade name Bollgard) to produce boll weevil–resistant cotton plants and into corn plants to resist the corn borer. This concept has been proven to work only in combination with other IPM techniques when infestations are extensive. Several other spore-forming bacteria (*Bacillus sphaericus* and *B. popillae*) also produce toxins effective against some insects. Finally, nuclear polyhedrosis, a baculovirus, has been shown to be an insect pathogen which has actually received EPA approval as an insecticide.

There is no question that the rapidly emerging biotechnology field will play a major role in developing ever more innovative IPM strategies for the future. However, a sobering note was introduced in 1999 when it was reported that Monarch butterflies (certainly a nontarget species) were also killed by corn plants grown from Bt treated seeds. Insects have a multimillion-year head start over humans on the evolutionary scale but human ingenuity remains a formidable force; the race is far from over.

## REFERENCES

Several references addressing issues of integrated pest management include:

Collins, F. H. and N. J. Besansky. "Vector Biology and the Control of Malaria in Africa." *Science,* 264: 1874-1875, June 24, 1994.

Office of Technology Assessment. 1995. Biologically Based Technologies for Pest Control. Report OTA-ENV-236. U.S. Government Printing Office. Washington, D.C.

# Section VI

# FUTURE CONSIDERATIONS

# Chapter 16

# ENVIRONMENTAL HEALTH IN THE TWENTY-FIRST CENTURY: A PEEK INTO THE FUTURE

Somewhere between cheerful dolts and nervous worrywarts there's a state of mind we ought to embrace.

— Carl Sagan, *Billions and Billions*

Predicting the future is risky business at best, and any such prediction based on the assumption that current trends will continue very far into a new century, let alone a new millennium, is even less certain. When growth is exponential (e.g., human population increase since the initial emergence of *Homo sapiens*), continuing such a trend also defies the laws of physics. Using the example of Malthusian population explosion, long before the next millennium ends, planet Earth would consist of a throbbing mass of protoplasm expanding at the speed of light. Thus, I had best confine my peek into the future only to the early years of the twenty-first century. As a teacher of environmental health for most of the last half of the twentieth century, I am in the twilight of my career and will not have to answer for the folly of my predictions for very long. Thus I will attempt to summarize this turn-of-the-century perspective by projecting environmental health trends beyond (but not very far beyond) the dawn of the new millennium.

Phenomenal growth in the twentieth century has not been confined to human numbers. Technological advances have also been phenomenal, and where these advances take us will profoundly influence the future of the species. However, most advances in technology currently bypass a high percentage of the world's disadvantaged population, and the social and economic future of the third world will have a major impact on the stability of world order needed to allow technology to flourish. Terrorists and rogue nations with access to weapons of mass destruction could disrupt the steady advance of technology for all.

While humans are the first and only species potentially capable of disrupting the balance of nature, nature itself remains a potent and sometimes unpredictable force.

From volcanic eruptions to devastating earthquakes, hurricanes, typhoons, floods, and tidal waves, humans cannot always control their own destiny. In the longer term, ice ages and global warming will cycle themselves regardless of the actions of humans. Although the probability may be small, a major meteor impact, such as the one that apparently doomed the dinosaurs 60 million years ago, could also potentially eliminate the human species.

The point is that even the most sophisticated of models and the most powerful computers are based on human input, which must include assumptions. The range of possible assumptions is obviously great enough to yield conclusions ranging from destruction of all life on earth to some unimaginable Nirvana of perpetual peace, prosperity, and good health for all. Both extremes are of course unlikely. Thus, I will confine my view of the future to more probable trends and possibilities within a smaller range of dynamic change.

## PREPARING FOR THE FUTURE

Given the above assumptions, the constructive approach for environmental health professionals is to understand the impact of the environment on human health, to communicate relative risk better than we have to the public and to aggressively seek political support for the public health infrastructure. Public health has always been difficult to explain to the public. It is least visible when it is most successful—that is, when epidemics are not occurring and the preventive strategies are working. Politicians must be made to understand that it is more expensive to have to gear up in order to deal with emerging human health problems than it is to maintain the preventive infrastructure already in place. For example, many blame the reemergence of tuberculosis in the 1980s on the dismantling of effective public health measures (such as supervised drug regimens for indigent patients). These could have been maintained at a fraction of the cost of dealing with the consequences. I believe that a united, well-trained, and politically savvy corps of professional public health workers is a major necessity if we are to have any chance of influencing the uncertain future ahead.

In an essay on the future of environmental health that I prepared in 1997, I suggested that optimism for the future was possible if public health professionals understood the past which got us to this point, developed a realistic vision of future goals, and formulated a plan of action to achieve those goals. Clearly, any optimistic forecast could be derailed by deterioration of world stability to the point where nuclear holocaust reared its head. Humanity could succumb to a nuclear winter that relegates humans to a science fiction type of primitive existence in competition with the cockroaches. I will assume that such a scenario will not happen.

In the remainder of this chapter I will attempt to lay out a vision for the future, addressing the array of environmental issues discussed in this book. I base my predictions on a scenario where terrorism is contained, wars are limited to the brushfire variety, and world population growth stays within the confines of capacity to sustain a stable society. Admittedly, these are optimistic assumptions, but they are within the realm of human possibility. Technology is an enormous ally. We need only to muster the will to advance our social and political institutions sufficiently to allow technology's benefits to be visited on all of us more equitably. We can enhance our ability to fight it out with nature if we don't have to fight it out with ourselves at the same time.

## A PEEK INTO THE FUTURE

### Global Warming

In my view, natural cyclical warming and cooling of the Earth over hundreds or thousands of years make short-term variations difficult to evaluate. Thus, our vision should not be clouded by the record warmth of the 1990s. Indeed, technology, to date, has not afforded us the capacity to alter weather very significantly. On the other hand, uncontrolled growth of greenhouse gas emissions is not sound environmental policy. In the long run, fossil fuel reserves will be depleted and alternative energy sources must be sought, global warming or not. Our goal should include reductions in atmospheric greenhouse gasses consistent with economic growth and with reasonable allowances for developing nations to industrialize.

I predict that renewable, less polluting energy sources will gradually replace fossil fuels while technological advances will enable us to improve end-of-pipe reduction efficiency. Technological advances will be based more on less polluting, more energy efficient, microchip systems. Global climate may indeed change during the twenty-first century. (I would not object at all if Minnesota became a few degrees warmer in January and February.) I predict, however, that the century will see neither a catastrophic rise in ocean levels nor diminution of agricultural capacity. We may see record high and record low temperatures. There will be El Niños, La Niñas and devastating natural weather catastrophes, but probably no more than in other centuries. Awareness and reasonable limitations on greenhouse gas emissions are called for. Panic is not.

### Ozone Layer Depletion

As I pointed out in Chapter 1, ozone layer depletion is one of the less controversial environmental health issues of the day. Measurements clearly indicate that seasonal ozone holes have occurred. The role of chlorinated fluorocarbons (CFCs) in destroying stratospheric ozone is generally agreed upon by scientists, and the health hazards of ultraviolet radiation, which is blocked by ozone, are also well documented. At this point, however, CFCs have been greatly restricted, and substitute technologies for refrigerants are in place. Thus, both the goal and the prediction for the future seem clear-cut: Stay the course, restrict the release of CFCs, and the ozone layer will eventually stabilize, thus minimizing related human health concerns.

## Fresh Water Resources

Fresh water is the wellspring of all life, and its availability is what sets this planet apart from the rest of the known universe as a haven for the diverse living creatures which share its bounty. Our challenge is to distribute that limited but renewable supply to all human communities, keeping it free from biological and chemical contaminants which adversely affect human health. Large-scale outbreaks of waterborne cryptosporidiosis, in the United States and other developed countries, have served to remind us that we have to pay more attention to our infrastructure. Cryptosporidium is a typical emerging pathogen, one of numerous such infectious agents that pose major challenges for the new century. I do not doubt that other previously unrecognized pathogens will appear periodically. New methods for rapid detection and fingerprinting of those agents will help in their control, but modernization of water treatment facilities and development of new disinfectants and/or filters effective against protozoan agents must be a priority. In developing countries, creation of a water and wastewater infrastructure will be a necessary step if progress is to be made in improving the quality of life for the billions of people now without access to a dependable supply of potable water or systems for wastewater disposal.

## Air Pollution

Like potable water, breathable air should be a birthright for all humans. Whether the obvious industrial pollutants, the serious smog caused by intentional burning to clear land for agriculture in Southeast Asia and South America, or the particulates in indoor cooking fires which kill millions of children in developing countries, air pollution is largely a creation of human activity. The obvious goal is to reduce such pollution to levels protective of human health. In industrialized countries I see continued progress in reducing pollution. The outlook for the developing world is less optimistic. Only through economic improvement (tied to social and political stability) can progress be made. It is ironic that the air pollution effects most damaging to human health have shifted to the least industrialized nations. If economic development is the key to prevention in those countries, then the whole cycle of controlling industrial emissions will need to be repeated.

In the United States, burgeoning traffic congestion in metropolitan areas looms as the biggest air pollution headache. Fueled primarily by relatively inexpensive gasoline and the trend toward gas-guzzling trucks and sport utility vehicles (SUVs) I see that problem getting worse before it gets better. Ultimately, another energy crunch, with an ensuing surge in the price of gasoline, may be needed to reverse the trend. I also predict that mass transit will ultimately become necessary at least in cities where congestion becomes unbearable. I foresee that breakthroughs in mass transit efficiency and convenience will make that option more palatable to an American public addicted to the independence of their private vehicles. I do not foresee that elected legislators in the United States will ever muster the will to impose the taxes on fuels which serve to reduce private vehicle use elsewhere in the industrialized world. Meanwhile, the pattern of more and more distant exurban developments with multiple commuting routes works strongly against both mass transit and any rise in fuel taxes.

### Waste Management

The continued profligate habits of Americans to consume and discard merchandise, the decreasing availability of suitable land disposal sites, and increasing NIMBY attitudes against municipal waste incinerators all serve to exacerbate the problem of solid waste management in the United States. Achieving the goals of waste minimization and greater recycling efforts will help to contain that problem in the twenty-first century. I predict that the current practical ceiling of 50 percent as an achievable recycling goal for a community will increase to 75 percent early in the next century. Waste minimization programs will serve to reduce the per capita waste generation figure by about 25 percent (from 4 pounds per day to 3 pounds per day). I also see an increased role for refuse derived fuel (RDF) plants to convert waste to electric power. All in all, an optimistic outlook for this area of environmental health.

### Nuclear Power and Radioactive Waste

We are obviously at a crossroads relative to the future of nuclear power in the United States. Until such time as fusion power is available (I do not see this happening until about the middle of the twenty-first century), continued generation of nuclear energy and its role in our energy future will depend on finding a permanent solution to the problem of high-level nuclear waste containment and storage. The Yucca Mountain, Nevada, site remains the most likely, if the U.S. Congress can muster the political will to dictate that solution. If not, another feasible site will have to emerge soon as a contender or nuclear power will cease, at least temporarily, to be a player in the power generation industry. From this it is clear that future predictions in one area (e.g., air pollution) are dependent on problem resolution in another area (e.g., nuclear power). If nuclear plants are shut down because there is no waste containment site, power generation will depend more on fossil fuels, including the most polluting coal sources, and air pollution problems may increase. At the same time, renewable energy sources such as wind, water, and solar power should continue to emerge as important sources of energy, as technological advances render them more dependable and economically competitive.

### Housing and Health

As discussed in Chapter 8, inadequate housing for the poor looms as a major issue at the turn of the century, with health and safety being only part of the problem. The goal is to assure that all Americans, regardless of economic status, have a place to live, free from inherent health and safety defects. Ideally, the free market approach should provide for that goal. If there is a strong economy, the housing industry should be able to provide reasonable-quality housing at an affordable price still high enough to make a profit for the builder or rental property owner. Current conditions, even in the economic good times of the late 1990s, do not seem to validate that concept. Public subsidies are clearly necessary, but, in the political reality of the times, are not adequately forthcoming. Exacerbated by the ongoing debate over requiring all communities to share the burden, rather than concentrating the poor in the inner cities, I do not see a solution near at hand. This remains one of my more pessimistic predictions. Unless some truly innovative mechanism can be found to house the poor decently but economically, I see this situation getting worse in the early years of the twenty-first century before it gets better.

### Occupational Health

Protecting worker health and safety would seem to be a goal with few detractors. However, in practice the future of that goal is tied to the ongoing struggle between labor and management. The outcome of conservative efforts to diminish the authority of OSHA and the future viability of labor unions will be determining factors in that struggle. Any prediction on the future direction of occupational health essentially becomes a prediction about whether conservatives or liberals control the political climate of the twenty-first century. If history is a guide, the pendulum will swing back and forth for the foreseeable future. It is also clear that this is not an all-or-nothing issue. Occupational health issues will see progress and setbacks regardless of political dominance. It is more a matter of degree than of absolutes. My prediction is for a long-term standoff with a gradual improvement in worker health (based mainly on technical advances), regardless of political outcome.

### Nosocomial Infections

One of the more frightening scenarios developing in the late 1990s is the capacity of pathogenic microbes to develop resistance to the array of antibiotics available to combat them. This problem is most apparent in the healthcare setting, where immunocompromised patients are concentrated. Part of the problem is the evolutionary advantage of microbes to select for resistance qualities quickly and efficiently. The problem has no doubt also been exacerbated by overreliance on antibiotics by the medical profession; too many doctors prescribe broad-spectrum antibiotics in the absence of definitive identification of the pathogen or knowledge of the most effective therapy.

The pharmaceutical industry must also share the blame for pushing high-profit antibiotics even in third world countries without thought to the consequences of resistance development. It is more important, however, to seek solutions than to seek scapegoats. Restrictions on antibiotic use, supervised regimens to assure that the full course of antibiotics is taken by the patient, and improved surveillance to detect resistance patterns are all necessary steps. My prediction is that new pharmacological approaches to pathogen control will be developed. For example, modifications at the DNA level to trick pathogens into not producing toxins are on the drawing boards. From an environmental health perspective, it is also my contention that we must return to the basic sanitation emphasis employed before modern antibiotics were available. More attention to handwashing, sterilization, disinfection, and general aseptic technique is recommended and predicted, to prevent resistant pathogens from gaining a foothold in healthcare environments. My prediction is that new pharmacological breakthroughs, together with a return to the basic precepts of sanitation, will enable us to keep up with emerging drug-resistance of pathogens, but the seriousness of this problem should not be underestimated.

## Food Safety

In the last two decades of the twentieth century, we witnessed food-borne illness in the United States from emerging pathogens such as *E. coli* 0157:H7 and cyclospora. We have seen outbreaks associated with previously recognized pathogens spread nationwide as food processing, distribution, and consumption patterns have changed. Extended international trade has exposed us to ever evolving opportunities for importing food with suspect history. We have responded with tightened regulations and reorganized surveillance and inspection systems. Our goal is to retain confidence that food can be consumed safely regardless of where it is produced, purchased, processed, or prepared. Reaching that goal will require strict enforcement of safety standards, applied equally to domestic and foreign sources. It will also require better risk communication and consumer education. I predict that we will eventually overcome consumer reluctance to accept innovation. Just as pasteurization of dairy products was opposed when first introduced, then widely accepted, irradiation of food, such as ground beef, which is susceptible to pathogen contamination, will eventually become commonplace. Similarly, fear of biotechnology, such as BST in milk, will also fade as the concept becomes familiar and anticipated hazards fail to materialize. Once again, I am optimistic. If we continue to support the oversight infrastructure, technological innovation and improved risk communication will lead to restored confidence in the safety of our food.

## Integrated Pest Management (IPM)

The IPM concept gained wide acceptance as "panacea" chemical pesticides were recognized as potentially more damaging than helpful. The emergence of new vector-borne pathogens has accentuated the need to continue and strengthen the IPM approach. Some of the factors contributing to the emergence are of human origin. Certainly, human intrusion into rain forests, unrestricted clearing of land for agriculture, and thoughtless exploitation of ecological environments without consideration of natural interactions must be stopped. The potential for innovation in IPM is unlimited. The new tools of biotechnology and new understanding of complex ecological niches involving many arthropod and mammalian species open the door to developing targeted approaches independent of damaging chemical pesticides. I predict that the twenty-first century will see enormous progress in such innovation. Once again I offer an optimistic scenario that human intelligence can prevail over even the most adaptable and persistent of disease carrying enemies.

## WHAT DOES IT ALL MEAN TO LIFE IN THE 21ST CENTURY?

In this chapter I have laid out a mostly optimistic view of human health and the environment in the twenty-first century. I do not mean to downplay the self-inflicted and natural dangers along the path we must travel. It is a perilous path indeed, strewn with potholes and sharp curves with precipitous drop-offs on all sides. I base my optimism on the many plusses which I perceive in us as humans. I believe that humans collectively are decent, compassionate, intelligent, and innovative. We could be on the brink of technological breakthroughs which will dwarf the incredible advances of the twentieth century. On the downside, there are those among us who are ignorant, arrogant, greedy, and shortsighted. For technology to flourish, we must come to grips with the social and political failings that have created an unsupportable chasm between the haves and have-nots, at home and abroad. The stakes are high. If we are to leave a legacy of hope and happiness to future generations, we must come up with a workable formula to define and achieve optimal "net societal benefit" in all of our actions.

# INDEX